R. L. Isaacson
Department of Psychology
University of Florida
Gainesville, Florida 32601

D1515367

Studies on the Piriform Lobe

Studies on the Piriform Lobe

F. VALVERDE, M.D. INSTITUTO CAJAL, MADRID

DEPARTMENT OF ANATOMY, HARVARD MEDICAL SCHOOL, BOSTON, MASSACHUSETTS

HARVARD UNIVERSITY PRESS · CAMBRIDGE, MASSACHUSETTS · 1965

Distributed in Great Britain by Oxford University Press, London

Library of Congress Catalog Card Number 65–16689

Printed in the United States of America

Publication of this book has been aided by a grant from the Commonwealth Fund

To the memory of Santiago Ramón y Cajal

Preface

The present studies have been made in an attempt to shed more light on the complex anatomical organization of the rhinencephalon. The author hopes that it will be of interest to neuroanatomists and neurophysiologists engaged in studies on this field.

The purpose of this investigation was firstly to study the connections of the amygdaloid complex by combining the silver-impregnation method of Nauta and Gygax and the Golgi technique, but the work grew out from the need to understand the organization and reciprocal connections of other brain structures with the amygdaloid complex. Thus, further lesions were made in other regions and the study of the Golgi material was extended to include the olfactory bulb, anterior olfactory regions, and piriform cortex. The accumulation of material justified the idea of writing a single monograph rather than a series of papers.

I am aware that this monograph is incomplete in many places, since it represents only the description of my experience and personal points of view. However I hope that it will represent at least one step toward a better understanding of the brain machine.

This work began in the fall of 1960 in the Instituto Cajal, Madrid. It was continued by the support of a grant from the Juan March Foundation, Spain, and finished in the United States of America in the Department of Anatomy, Harvard Medical School, Boston, with the aid of grant FF 515-S1 of the United States Public Health Service. These aids are acknowledged with thanks.

I am deeply grateful to Dr. A. Fernández de Molina, Instituto Cajal, who was the promoter of these investigations, and whose continuous help and encouragement were of great value for the accomplishment of this work; to Dr. Sanford L. Palay, Department of Anatomy, Harvard Medical School, and Dr. Walle J. H. Nauta, Massachusetts Institute of Technology, for many suggestions; to Dr. Elizabeth Taber, Department of Anatomy, Harvard Medical School, for her kind assistance and great care in improving the English; to Mr. Antonio de la Fuente, Instituto Cajal, for the construction of the coagulation device with which the experimental lesions were made and for technical assistance during some of the experiments; to Mr. Leo J. Talbert, Harvard Medical School, for his help in making the photomicrographs which illustrate the results obtained with the triple-impregnation procedure of the Golgi method; and to Mrs. Phoebe Franklin, Department of Anatomy, Harvard Medical School, who has taken care of the typing of the manuscript.

Acknowledgment is made to the editors of the *Journal of Anatomy,* London, and *Trabajos del Instituto Cajal de Investigaciones Biológicas,* Madrid, and to the Elsevier Publishing Company, Amsterdam, for permission to reproduce some figures from my own previous publications.

F. Valverde

Boston, Massachusetts
June, 1964

Contents

Studies on the Piriform Lobe

Introduction

One of the most interesting spectacles in phylogeny is presented by the evolution of the olfactory brain from lower mammals to man. The olfactory brain, the principal sensory implement with which the lower forms face their environment, has been reduced in man to a group of rudimentary formations. Undoubtedly, olfactory functions have been attributed to parts of the brain which, even in macrosmatic animals such as the cat, may be only loosely related to the sense of smell. The amygdaloid complex is an example of such structures. What prevents us from thinking that higher nervous elaboration, strange to any olfactory sensation, develops at glomerular levels, for example, in the olfactory bulb? Concerted efforts along parallel anatomic, physiologic, and behavioral lines are needed to provide as much as a glimpse into the intricacy of even the most primordial brain.

Broca (1878) gave the name of "great limbic lobe" to those regions of the brain surrounding the hilus of the hemisphere; he pointed out that the limbic lobe forms a common denominator in the mammalian brain, a point of view supported later by G. E. Smith (1910). Concerning comparative anatomy much information has been gained through the studies of, among others, Cajal (1911), Johnston (1911, 1923), Herrick (1924a, 1924b, 1933), Crosby and Humphrey (1939, 1941, 1944), and Koikegami (1963b).

The limbic system includes some subcortical cellular groups, all of which play an important role in emotional behavior. Bard (1928), Cannon (1929), Wheatley (1944), and Hess (1949, 1954) pointed out the importance of the hypothalamus as a center for emotional behavior. Papez (1937) proposed that a series of neuronal interconnections relating the entorhinal and hippocampal regions with the thalamus and hypothalamus may provide the anatomical substrates for the mechanism of emotion. This hypothesis was later extended by MacLean (1949, 1955a, 1958), who analyzed emotional behavior along the lines of modern psychosomatics (MacLean, 1960). Even the mesencephalon has been regarded as an important center (the limbic midbrain area of Nauta, 1958) for correlation between the limbic regions and other subcortical zones, but, among these subcortical centers, the amygdaloid complex has attracted much attention in view of the complex behavioral changes associated with its stimulation or ablation. Although earlier studies by Brown and Schäfer (1889) showed certain specific patterns of behavior following ablation of parts of the rhinencephalon including the amygdala, it was during the last fifteen years that several experimental studies revealed that the amygdala is a subcortical center providing a wide variety of responses. Kaada (1951), Koikegami and Fuse (1952), Koikegami, Kushiro, and Kimoto (1952), Koikegami and Yoshida (1953), Koikegami, Kimoto, and Kido (1953), MacLean and Delgado (1953), Kaada, Andersen, and Jansen (1954), Koikegami, Yamada, and Usui (1954), Magnus and Lammers (1956), Shealy and Peele (1957), and Fernández de Molina and Hunsperger (1959) obtained complex autonomic responses following stimulation of the amygdala. Gastaut *et al.* (1952), and Baldwin, Frost, and Wood (1954, 1956) obtained patterns of face and eye movements after stimulation in the amygdala. In man the stimulation of the amygdaloid complex produced the sensation of intense fear (Heath, Monroe, and Mickle, 1955).

Animals with ablations that included the amygdala of both sides as well as adjacent parts of the rhinencephalon showed behavioral patterns consisting of hyperactivity, loss of fear, and aberrant sexual behavior among other symptoms (Klüver and Bucy, 1939; Pool, 1954; Pribram and Bagshaw, 1953; Schreiner and Kling, 1953). It is interesting to point out that the "amygdala syndrome" is abolished by destruction of the ventromedial hypothalamic nuclei (Kling and Hutt, 1958).

The participation of the amygdala in emotional stress has been studied by Bovard and Gloor (1961), among others.

Other kinds of experimental work have shown that the amygdala represents a part of the subcortical neural system controlling defensive behavior (Fernández de Molina and Hunsperger, 1959, 1962; Hilton and Zbrożyna, 1963).

The importance of the amygdala was early recognized by Jackson (1889), who described a case of epilepsy originating in this region, thus initiating the development of one of the most interesting chapters of neuropathology: the study of psychomotor epilepsy.

From the preceding notes one obtains the idea that the amygdala has stimulated a great deal of thought concerning its functions, even though there exists, surprisingly, a considerable lack of anatomical investigations in this field. Much work has been done along physiologic lines. The amygdala, placed in an intermediate position between the cortex and the diencephalon, shows so wide a range of responses that "its stimulation suggests that there is a peculiar and widespread pattern of projection from this structure into the subcortical areas governing neuronal mechanisms which influence global activities of the organism as a 'whole' . . . The amygdala may be a structure, among others, which gives a high degree of flexibility of function to the nervous system" (Gloor, 1955a).

Recent electrophysiologic studies tend to demonstrate the great extent of influence of the cortex entorhinalis, gyrus cinguli (the mesopallium of MacLean, 1955b), and hippocampal formation (archipallium of MacLean, 1955b) over the ascending stream of neural activity of the reticular

formation of the diencephalon and brain stem. The existence of such a powerful influence, supported by the experimental observations of Adey (1958) and many others, would be sufficient to emphasize the importance of the rhinencephalon from the point of view of the recent neurophysiological concepts.

The present investigation does not intend to be a comprehensive study of the olfactory brain. For complete reviews of the amygdala and rhinencephalon the reader should refer to the works of Gloor (1960), Gastaut and Lammers (1961), and Koikegami (1963b), and thus, although studies on the olfactory bulb, the cortex piriformis, and certain regions of the diencephalon have been included, the present studies remained centered mainly in and around the amygdaloid complex and its related field. Part of this study has been published as short papers presented at the Third International Meeting of Neurobiologists held at Kiel, Germany, in September 1962 (Valverde, 1963c), and at the 76th and 77th Annual Meetings of the American Association of Anatomists (Valverde, 1963b, 1964a, b, c). Some observations have been published (Valverde, 1962, 1963a).

The classification given below is intended to delimit, from an anatomic point of view, the olfactory parts of the brain from other regions of the "visceral brain." This classification is based on the present observations and on the classifications proposed by Thomalske, Klingler, and Woringer (1957) and Gastaut and Lammers (1961). It includes, among other structures, the nucleus medialis dorsalis of the thalamus and the gyri orbitalis and sylvianus posterior of the cat on account of the connections existing between these structures and the amygdaloid region.

I. Olfactory brain *in sensu strictu*
  A. Area bulbaris
      a. Bulbus olfactorius
      b. Bulbus olfactorius accessorius
  B. Direct olfactory projection area
      a. Area retrobulbaris, Pedunculus olfactorius, Nucleus olfactorius anterior
      b. Gyrus and tractus olfactorius medialis
      c. Regio tractus olfactorii lateralis

1. Tractus olfactorius lateralis
2. Tuberculum olfactorium
3. Nucleus tractus olfactorii lateralis
4. Area praepyriformis
5. Area amygdaloidea anterior
6. Nucleus amygdaloideus corticalis
7. Area periamygdalaris

II. Lobus limbicus
   A. Area adolfactoria, Area subcallosa
   B. Gyrus cinguli
   C. Gyrus hippocampi, Gyrus parahippo-campalis, Cortex entorhinalis
   D. Uncus

III. Formatio hippocampalis

   A. Cornu Ammonis ⎱ including the
   B. Gyrus dentatus ⎰ rudimentary pericallosal formations
      a. Fasciola cinerea
      b. Flexura subsplenialis
      c. Flexura retrosplenialis
      d. Indusium griseum
         1. Stria longitudinalis lateralis
         2. Stria longitudinalis medialis
      e. Gyrus subcallosus, Gyrus paratermi-nalis
   C. Limbus Giacomini
   D. Gyrus fasciolaris

IV. Subcortical formations related mainly with the olfactory brain
   A. Gyrus diagonalis, Lemniscus diagonalis (Snider and Niemer, 1961), Diagonal band of Broca
   B. Regio praeoptica
   C. Bed nuclei of the stria terminalis and commissura anterior
   D. Nucleus ventromedialis hypothalami
   E. Hypothalamus lateralis including the fasciculus prosencephali medialis (Snider and Niemer, 1961) or medial forebrain bundle
   F. Area septalis
   G. Corpora mammillaria
   H. Nucleus anteroventralis thalami
   I. Nucleus medialis dorsalis thalami
   J. Nucleus habenulae
   K. Nucleus amygdalae excluding the nuclei amygdaloideus corticalis and tractus olfactorii lateralis

V. Région périvalléculaire or périfalciforme of Gastaut and Lammers (1961)

Cat ⎰ A. Gyrus orbitalis
    ⎱ B. Gyrus sylvianus anterior ⎱ Primate
      C. Gyrus sylvianus posterior ⎰ insula

Olfactory impulses coursing through the olfactory nerves make contact with the descending dendrites of the mitral, periglomerular, and tufted cells of the olfactory bulb at the olfactory glomeruli. The mitral axons convey impulses throughout both tracti olfactorius medialis and lateralis toward the main receptive olfactory-brain areas represented by the nucleus olfactorius anterior, the gyrus olfactorius medialis, and the region of the tractus olfactorius lateralis comprising part of the tuberculum olfactorium, the nucleus tractus olfactorii lateralis, the nucleus amygdaloideus corticalis, the area amygdaloidea anterior, and the prepiriform and periamygdaloid cortices.

From the direct olfactory projection area, impulses are relayed through more or less complicated neuronal chains to the lobus limbicus, the formatio hippocampalis, and several subcortical and neocortical regions wherein higher nervous mechanisms relate to each other, integrate, and influence practically all the sensory modalities, exerting powerful influences on the reticular activating ascending systems.

1   Material and Methods

The present work is based on the observations made on a total of 81 brains, subdivided as follows:

Experimental (cats)	25
Golgi method (rats, mice, and kittens)	46
Heidenhain method, control (cats)	10

EXPERIMENTAL PROCEDURE

Adult, apparently healthy, male and female cats ranging between 1800 and 2800 g in weight were anesthetized intraperitoneally with Thiopenton Sodium. Lesions were made in the brain by means of high-frequency coagulation. A current of 20 to 38 ma for 10 sec was passed through a fine insulated needle with a 1-mm bare tip that was lowered into the brain by means of a stereotaxic head holder.

The animals were allowed to survive for 6 to 8 days after coagulation, and then, under ether anesthesia, fixation of the brain was initiated by perfusion with normal saline followed by 10-percent neutral formalin. The brain was removed and placed in 10-percent neutral formalin for 2 to 3 months. Pairs of adjacent frozen sections 20 μ thick were selected. One of each pair was stained by the Heidenhain method, the other by the Nauta and Gygax (1954) silver technique employing the Laidlaw's solution described by Albrecht and Fernstrom (1959). Slight modifications of the Nauta and Gygax method were adopted in order to obtain better contrast of the degenerating fibers and to afford easier identification of nuclear groups in the same stained Nauta section. The modification consists in a complementary hematoxylin staining of each section after gold chloride treatment. The procedure is as follows:

1. After treatment in the reducing Nauta solution, the sections are washed and immersed for 1 to 3 min in a 1/600 yellow gold chloride solution.
2. The sections are washed and transferred for 3 min to a 5-percent sodium thiosulfate solution.
3. After washing the sections are immersed in Carazzi's hematoxylin[1] for 1 to 2 min.
4. The sections are washed, dehydrated, and mounted by means of creosote.

The Heidenhain section of each of the chosen pair was projected on the paper and sketched in order to obtain an anatomical drawing of the plan; afterward the corresponding Nauta section, treated by the hematoxylin, was projected on the previous drawing to record axon degeneration. In the drawings the identification of the brain structures is based on the Jasper and Ajmone-Marsan (1954) and Snider and Niemer (1961) atlases with approximate adjustment of millimeter references. Coarse dots represent fibers of passage, fine stipple preterminal and terminal ones. The lesions are indicated in solid black.

THE GOLGI METHOD

For almost a century, the "black reaction" of Golgi (1873), obtained when pieces of brain tissue are silvered after treatment by an osmic-dichromate mixture, has been one of the best methods

---

[1]Carazzi's hematoxylin is prepared as follows: distilled water, 400 ml; glycerin, 100 ml; potassium alum, 25 gm; potassium iodate, 0.1 gm; hematoxylin, 0.5 gm. The solution should be exposed to the light and air for two weeks before use.

for the demonstration and study of the structure of the nervous system. The Golgi technique has been shown to have several advantages over other impregnation methods. For example, there are relatively few impregnated cells in the sections, and the transparent and yellowish background that occurs permits the obtainment of sections 200 to 300 μ in thickness. In these sections, if the orientation has been fortunate, very long axonal trajectories can be readily studied. Thus, in sagittal sections of mouse brain, I have been able to follow a single and uninterrupted axon from its origin in cortical pyramidal cells to the lower brain-stem levels. Further, the staining of relatively few cells and fibers in each section, the dendritic orientation, collateral axonal pattern, cell shape, and forms of fiber endings show most clearly.

It is disadvantageous to attempt to overstain sections by prolonging the time of treatment in osmic-dichromate or in silver, since precipitates often form at sites where fine fibers give rise to densely packed terminals, and so mask their delicate organization. In the past four years my efforts have been directed toward obtaining a more complete staining of all axonal collaterals and main branches rather than producing a stain that shows a greater number of nerve cells, for the osmic-dichromate Golgi method is not suitable for the study of cell clusters, dendritic interrelations, and comparative cell sizes. These are best studied by the Cox (1891) variant of the mercuric Golgi (1879, 1891) technique, although with this variant, usually known as the Golgi-Cox method, axons of cells cannot be followed very far beyond their origin, so that the terminal-fiber picture as revealed by the osmic-dichromate procedure is not obtained. The rapid Golgi method generally succeeds only in brains of immature animals, perhaps because myelin interferes in older animals. This idea has been tested recently (Valverde and Sidman, 1965) in mutant mice with deficient myelination. It has been found that these animals showed far more complete staining of cell bodies, processes, and terminal ramifications.

Three principal factors play a fundamental role in obtaining successful preparations: the age of the specimen, the size and form of the pieces used for impregnation, and the period of time elapsing between death and immersion in the osmic-dichromate mixture.

(*a*) *Age of specimen.* There are considerable variations in regard to the optimal age of a given specimen. According to my experience, the following are the ages at which different specimens give the best results with the osmic-dichromate reaction: chicken, any time in its embryonic development; mouse, 6 to 10 days; rat, 3 to 7 days; rabbit, 2 to 6 days; cat and dog, newborn to 3 days; man, embryos 5 to 8 months of menstrual age.

There is no doubt that acceptable results can be obtained by using specimens at ages other than these, but the results are not so good for the purpose of drawing a complete picture of the axonal trajectories and ramifications; if younger specimens are employed the dendritic organization and shape of the cell body will differ from that of the adult animal and some of the axonal collaterals will be only beginning to form from their parent main axonal branch. Also, when older specimens are employed, considerable parts of the axons, which are covered by the myelin sheaths, remain unstained. Unfortunately, with the use of specimens at the ages given above, the dendritic morphology shows considerable variation from that of the adult animal, especially regarding the dendritic thorns, which are almost completely absent at these ages.

(*b*) *Size and form.* Cubes of tissue 4–5 mm on an edge give the best results. Such blocks of tissue can be cut from a brain that has been removed and kept in a Petri dish containing saline solution, using either sharp scissors or a razor blade. The entire brain stem of the mouse can be immersed in the osmic-dichromate mixture. The medulla oblongata of the young rat, cat, and dog should be divided into halves by either a sagittal or a horizontal cut. The medulla oblongata of a human embryo needs to be divided into at least 4 blocks. I get successful impregnations of the brains of 3-day-old rats by dividing each hemisphere into only a dorsal and a ventral half by a horizontal cut, but it should be borne in mind that smaller pieces give the most complete impregnations. The spinal cord can be impregnated as pieces 3 mm long or in 5- to 6-mm lengths when the cord has been divided into halves, either sagittally or horizontally.

When possible it is advantageous to keep the structure to be studied in the center of the block; for example, if one is trying to stain the raphe cellular groups of the medulla in the cat, the pieces should be delimited parasagittally by cuts passing through the trigeminal nuclei, for, if the medulla is divided midsagittally, the superficial layers of the block become completely penetrated by deposits of silver chromate so that it is impossible to identify any structure. For similar reasons, when possible the meninges should be left in place.

(*c*) *Freshness of the pieces.* No major difficulties occur with newly killed animals, since not more than 5 to 10 min should elapse between the removal of the brain and immersion of the pieces in the osmic-dichromate mixture. However, this is not possible with human material. Nevertheless, I have obtained successful results with pieces of brain taken from human embryos 8 hr after abortion. Human embryos about 7 to 8 weeks old taken 24 hr after death, from women who died in accidents, completely failed to show any stained nervous structure. Brain pieces of one human child who died 24 hr after birth, and which were immersed in the osmic-dichromate mixture 30 hours after death, showed a very poor reaction. As the conditions of preservation depend upon factors such as temperature and permanence of the embryonic envelopes, 10 hr is the limit of the period within which successful results can be obtained.

(*d*) *The procedure.*

A stock solution is prepared as follows:

Potassium dichromate	12 gm
Osmium tetroxide	1 gm
Distilled water	500 ml

To make this solution first dissolve the potassium dichromate in the distilled water. When its solution is complete, a sealed glass tube containing 1 gm of osmium tetroxide is carefully cleaned by removing the label and all vestiges of glue, dried, and kept on dry ice until the crystals of osmium tetroxide separate from the walls of the tube. Then the glass is broken in the middle and the crystals are dropped into the potassium dichromate solution. The solution should be complete in about 10 hr; it can be accelerated by using a magnetic stirrer. The osmic-dichromate mixture can be stored in a dark bottle with a glass stopper for 4 to 5 months in the refrigerator.

In addition, a 2-percent solution of osmium tetroxide is also prepared and kept in the refrigerator. As aqueous solutions of osmium tetroxide are very easily spoiled by reduction, this should be freshly prepared in small quantities.

(1) Immerse pieces of nervous tissue not thicker than 4 to 5 mm, taken from a recently killed animal, in the following:

Stock osmic-dichromate solution	20 ml per piece
2-percent osmium tetroxide	5 ml per piece

Keep the bottle or bottles containing the pieces for 7 days at room temperature.

(2) Wash the pieces briefly in a small container with 0.75-percent silver nitrate. A precipitate is formed. Then, put the pieces into a new 0.75-percent silver nitrate solution for 2 days. Use 100 ml of silver solution per piece and keep at room temperature in a dark place.

(3) Without washing, the pieces are transferred directly to a fresh osmic-dichromate stock solution, using 25 ml per piece. The pieces are left in this solution for 4 days at room temperature.

(4) Repeat as in (2) but leave the pieces in the silver solution for 3 days at room temperature. Use a new silver solution.

This completes the second impregnation; the pieces can now be tested by cutting 3 or 4 sections as described below. If the impregnation is incomplete or poor, the pieces should be subjected to a third impregnation as follows:

(5) The pieces are transferred directly to the following:

Fresh stock osmic-dichromate solution	25 ml per piece
Potassium dichromate	0.5 gm per piece

Keep the pieces in this solution for 4 days at room temperature.

(6) Repeat as in (2) but leave the pieces in the silver bath for 5 days. Use a new silver solution.

(7) Dehydrate the pieces in absolute alcohol for 10 to 20 min.

(8) Stick the piece of brain tissue in a cube of paraffin wax. To do this, a block of paraffin is trimmed so as to form a cube slightly larger than the piece of brain tissue. Fix the cube of paraffin in the object-holder of the microtome. With a hot needle, melt the paraffin of the upper surface to make a hole, then rapidly take the piece of tissue from the alcohol and press it gently into the hole in the melted paraffin. Before sticking the piece of brain tissue in the paraffin, ensure that the block is properly oriented. With the warmed needle, melt paraffin of the cube and build up a thin paraffin wall around the lower third of the piece in order to ensure a firm attachment.

During this operation drying of the tissue should be avoided. *Keep the tissues constantly wet with absolute alcohol,* using a small brush or a dropper. Finally, screw the object-holder with the piece into a sliding microtome. Use a planoconcave knife, adjusting it to form an angle of 10° between the knife and the block face.

(9) Cut sections at 100, 200, or 300 μ, keeping the upper surface of the knife constantly soaked with absolute alcohol. Use a razor blade and a fine brush or needle to transfer the sections through the following steps. Carefully remove the section from the knife by sliding the razor blade between the section and the knife.

(10) Transfer the sections to a Petri dish containing absolute alcohol for 30 min. Sections may be kept in sequence by flattening them against the bottom of the dish.

(11) Transfer each section individually to a Petri dish containing oil of cloves. Allow each section to float on the surface of the oil for a few seconds and then push it down with the brush into the bottom of the dish, maintaining the sequence. In transferring sections avoid folding. Keep the sections in the oil for 30 min.

(12) Remove the sections from the oil of cloves individually and mount them on a slide placed horizontally on the table near the dish. In order to prevent drying of the sections on the slide, cover it with a layer of oil of cloves before the sections are transferred.

(13) Incline the slide slightly and carefully drain the oil of cloves with a brush soaked in xylol. Continue washing with xylol until it runs clear.

(14) Drain the xylol and cover the sections with Damar resin, Canada balsam, or Permount. Allow to dry and afterward add other layers of resin until the sections become completely embedded in it. Do not cover the sections with a cover slip.

The present procedure, like a previous one (Valverde, 1962), was developed from that described by Cajal (1889) and Cajal and Castro (1933). In comparing the original preparations of Cajal (from the Cajal Museum at the Instituto Cajal, Madrid) with mine, I found (taking into account the fading of his preparations after almost half a century) that the Cajal procedure shows more stained cells than in my preparations (see Fig. 2, *B*), but that the terminal axonal pattern is not so clear and complete as in the preparations stained by the present procedure.

It is sometimes advantageous to stop the process of staining after the second impregnation, that is, to pass directly from step (4) to step (7). With the second impregnation one frequently finds more stained cells than after the third immersion in the osmic-dichromate mixture, but the axons and their collaterals are incompletely stained, as are the terminal ramifications. Sometimes it is best to keep the pieces in the refrigerator during the first impregnation; this has proved advantageous with human material which is usually in a poor condition of preservation.

Figures 1 and 2*A* were made with the intention of showing the results of the present procedure. The photomicrographs were taken with the following: objective, apochromat 10 ×; ocular, Homal No. 1; Wratten filter E; neutral-density filters varying from 0.3 to 1.2 according to the thickness of the section; film, Kodak Panatomic-X, 4 × 5 in.

In the Golgi drawings of the present work the cells were lettered; the main axonal branch of each cell was always labeled *1* followed by the letter of its parent cell; the collaterals of the axons were numbered consecutively, followed also by the letter of the parent cell. I hope that this will provide easy identification of the cell to which a given axon or branch belongs.

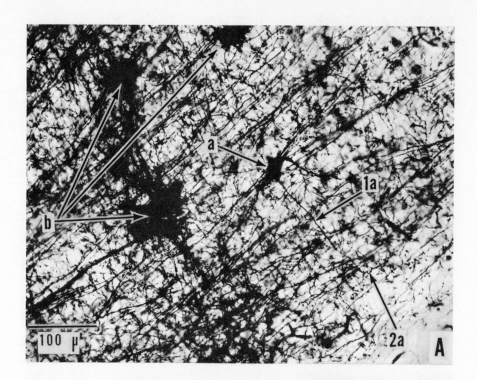

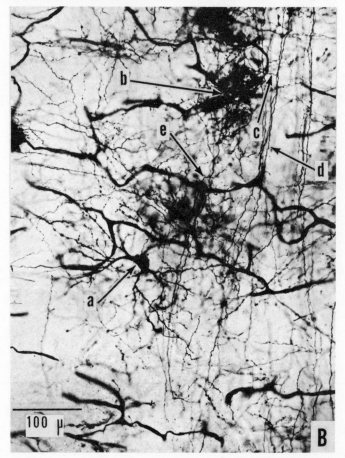

Fig. 1. (*A*) Parasagittal section of the medulla oblongata, showing a very complete staining of longitudinally running fibers and delicate terminal reticular neuropil. One cell (labeled *a*) of the nucleus reticularis gigantocellularis is slightly out of focus; its axon (*1a*) can be followed to the point marked by arrow *2a*, where it changes its course to run sagittally. Arrows *b* indicate three deposits of silver chromate. Rat, 3 days old. Golgi method.

(*B*) Parasagittal section of the medulla oblongata through the nuclei of posterior funiculus. One cell (*a*) of the nucleus gracilis has its axon out of focus. Arrow *b* points to the dense terminal ramifications of fibers *e* and *c*, the latter being a collateral of fiber *d*. Rat, 3 days old. Golgi method.

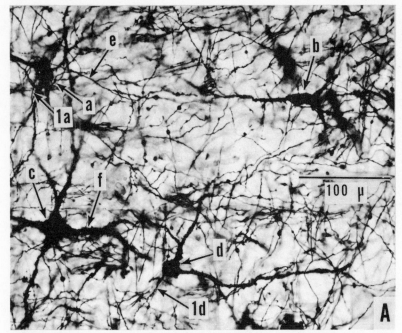

Fig. 2. (*A*) Transverse section through the medulla oblongata. Arrows *a*, *b*, *c*, and *d* point to four cells of the nucleus medullae oblongatae centralis, subnucleus dorsalis (Olszewski and Baxter, 1954). The axons of *a* and *d* are indicated by arrows *1a* and *1d* respectively; the axon of cell *c* is out of focus; the axon of cell *b* was not observed. Fiber *e* supplies fine terminal fibrils enveloping cell *a*. The initial part of the thick dendrite shown by arrow *f* is surrounded by many crowded synaptic end bulbs. Human embryo, 6 months old. Golgi method.

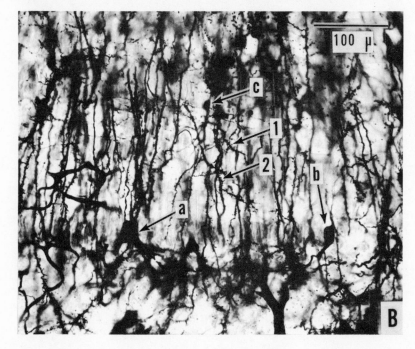

(*B*) Transverse section through the olfactory bulb. Arrows *a* and *b* point to two mitral cells, and arrow *c* to an internal granular cell. Some synaptic contacts on the descending expansion (axon) of the internal granular cells are still visible in the regions indicated by arrows *1* and *2*. This photomicrograph was taken from a preparation made by Cajal about 60 years ago. Cat, 25 days old. Golgi method.

2 Experimental Observations

The lesion of this animal (Fig. 3, sections 6 to 8 and Fig. 4, section 3) is confined to the anterior part of the amygdala. It comprises largely the rostral third of the pars medialis of the nucleus amygdaloideus centralis (A.c.m.), with involvement of the dorsal and caudal parts of the area amygdaloidea anterior (A.a.) where the latter fuses with the pars lateralis of the nucleus amygdaloideus centralis (A.c.l.; Fig. 3, section 8). The fasciculus longitudinalis associationis (F.l.a.) also was involved in the lesion. The needle tract had caused additional damage through the lateral part of the globus pallidus (G.P.).

As shown by Fig. 4, three main systems of degenerating fibers could be traced from the lesion: (*a*) a medial system, identified as the medial amygdalo-hypothalamic pathway; (*b*) a second system which extends rostromedially following the course of the fasciculus longitudinalis associationis; and (*c*) a third group spreading anteroventrally. No fiber degeneration was observed in the stria terminalis.

(*a*) The medial system (Fig. 4, section 3), or amygdalo-hypothalamic pathway, extends from the superior pole of the area of coagulation to the hypothalamus lateralis (H.L.) through the ansa lenticularis (An.len.). This system forms a group of fine and relatively condensed fibers lying just dorsal to the nucleus supraopticus (So.). It spreads out in the hypothalamus covering a zone extending medially to the hypothalamus anterior (H.A.), without, however, displaying signs of termination in the latter. The area in which the terminal degenerating fibers were identified covers a zone of the hypothalamus lateralis including the neighborhood of the ventral half of the pars descendens columnae fornicis and the medial forebrain bundle. Ventrally, fine degenerating fibers extend to the region above the tractus opticus (T.O.); no degeneration was traced to the nucleus supraopticus (So.). Rostrally, the limits of this projection field could not be identified because it overlaps with the degeneration of the regio praeoptica traced via the fasciculus longitudinalis associationis. Caudally, degeneration in the medial forebrain bundle through the hypothalamus lateralis does not appear to run far and only scanty terminal degeneration could be identified at premammillary levels.

The system of fibers just mentioned forms a component of the ansa lenticularis, the latter being a major system that contains fibers originating in the temporal cortex and other subcortical structures and courses dorsal to the amygdala underneath the globus pallidus to the hypothalamus and further. The fibers described here originate in the amygdala; however, some involvement of temporohypothalamic fibers by the lesion cannot be excluded.

(*b*) The second system of degenerating fibers courses through the fasciculus longitudinalis associationis (Fig. 4, section 2, F.l.a.). It spreads out medially underneath the nucleus ansae lenticularis (N.An.len.) in the regio praeoptica (R.P.O.) and the medial zone of the area amygdaloidea anterior (A.a.). As shown in section 2, the degenerated fasciculus longitudinalis associationis forms a condensed group of coarse fibers located in the limit separating the regio praeoptica from the area amygdaloidea anterior (see also Fig. 18*A*).

In this section (Fig. 4, section 2) the system has been cut obliquely as it approaches the regio praeoptica, where numerous degenerating fine fibers leave it and distribute themselves to the rostral half of this region and the dorsomedial zone of the area

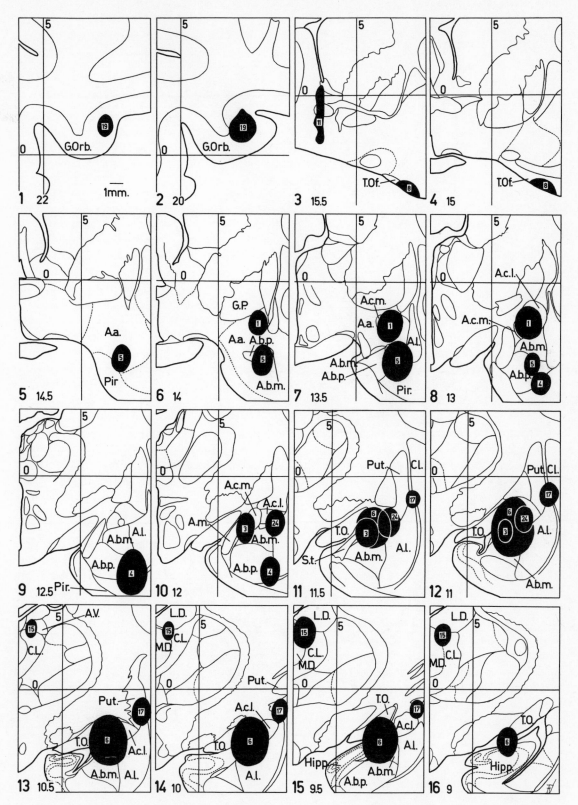

Fig. 3. Diagrammatic representation on frontal sections (Jasper and Ajmone-Marsan, 1954) of the different coagulations made in the present series of experimental cats. The lesions are indicated in solid black; the number of each lesion refers to the corresponding cat. Sections have been numbered consecutively from 1 to 16. For the meaning of the abbreviations used in this and other figures, see pp. 118–119.

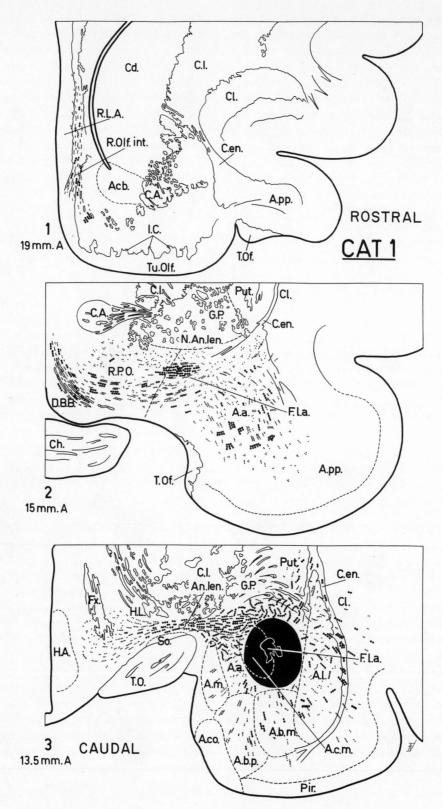

Fig. 4. Cat 1. Degeneration following lesion of the anterior part of the amygdaloid complex.

amygdaloidea anterior. The medial preoptic region is, however, free of terminal degenerating elements. The fiber degeneration traced through the fasciculus longitudinalis associationis to the area praeoptica is accompanied by several condensed groups of coarse as well as fine fibers which, coursing in the ventral half of the regio praeoptica, follow the diagonal band of Broca (section 2, D.B.B.). They continue upward through the radiatio olfactoria interna (section 1, R.Olf.int.) to the regio limbica anterior (R.L.A.). Fine degenerating terminal fibers ascend parallel to the midline, without penetrating, however, among the cells of this interhemispheric formation.

(c) A third, rather diffuse, system of degenerating fibers extends from the area of coagulation following rostral and ventral trajectories. From the ventral aspect of the lesion (Fig. 4, section 3) numerous groups of fine as well as coarse fibers extend in fantail fashion to the remaining amygdaloid nuclei, excluding the nucleus amygdaloideus corticalis (A.co.). From the lateral aspect of the lesion small fascicles of coarse fibers enter the capsula externa (C.en.). A few scattered degenerating elements can be seen in the deep layers of the cortex piriformis (Pir.).

Rostral to the lesion a moderate number of small fascicles have been cut transversely (section 2). These fascicles enter the area amygdaloidea anterior (A.a.), which is pervaded by numerous degenerating fine axons of terminal character. This lateral part of the rostral projection system of the amygdala appears to consist of a short-axoned cell core which is present in the area amygdaloidea anterior (see Golgi material and Fig. 35).

CAT 3

As shown in Fig. 3, sections 10 to 12, and Fig. 5, section 3, in this animal the lesion involved the medioventral part of the caudal half of the nucleus amygdaloideus centralis, pars medialis (A.c.m.); the caudal tip of the dorsal edge of the nucleus amygdaloideus medialis (A.m.) and part of the adjoining dorsal region of the nucleus amygdaloideus basalis, pars magnocellularis (A.b.m.). The medial extension of the central third of the pars lateralis of the nucleus amygdaloideus centralis (A.c.l.) also was included in the area of coag-

ulation. The lesion interrupted two main systems of fibers of the amygdala: the stria terminalis (S.t.), which was involved in the dorsal part of its origin in the amygdala (section 3), and the fasciculus longitudinalis associationis (F.l.a., section 2).

As shown in Fig. 5, five systems of degenerating fibers were traced from the area of coagulation: (a) the amygdalofugal part of the stria terminalis; (b) a medial system, similar to that described in Cat 1, identified as the amygdalo-hypothalamic fibers or the amygdaloid component of the ansa lenticularis; (c) a lateral system connecting the amygdala with the tip of the temporal pole; (d) rostrally, degenerating fibers through the fasciculus longitudinalis associationis; and finally (e) a rather diffuse system, which spreads out ventrolaterally, invading the other nuclei of the amygdala.

(a) The degeneration of the stria terminalis was followed to the commissural level (Fig. 5, section 2, S.t.), where fine terminal elements could be identified in the large expansion that forms the bed nucleus of the stria terminalis (B.S.t.). The latter is pervaded by numerous degenerating fibers. Coarse degenerating elements appear to bypass ventrally the commissura anterior (C.A.) and to distribute themselves in the dorsomedial part of the regio praeoptica (R.P.O.), where they overlap with the degeneration traced through the fasciculus longitudinalis associationis (F.l.a.). Consequently, the site of termination of neither tract in this area could be decided.

A group of fibers from the stria terminalis curves medially and descends toward the commissura anterior, entering the superficial stratum of the commissura, where fine degenerating elements could be traced throughout this stratum. Golgi material (see Fig. 32) shows that these intracommissural fibers synapse with dendrites of the surrounding gray matter that enter the commissura anterior.

(b) A moderate number of degenerating fibers stream medially from the lesion to the hypothalamus lateralis (Fig. 5, section 3, H.L.). Terminal degeneration is present in the medial forebrain bundle area (M.F.B.) and around the lateral aspect of the fornix (Fx.). This medial system courses in the subcapsular transit of the ansa len-

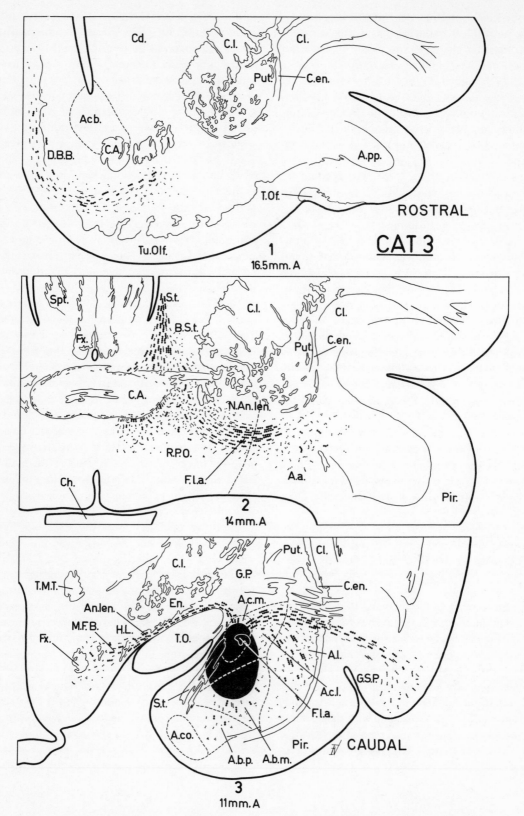

Fig. 5. Cat 3. Degeneration following lesion of the medial region of the amygdaloid complex.

ticularis (An.len.). In this case the terminal degeneration found in the hypothalamus was scarce and no degenerating elements were identified rostral or caudal to the area shown in section 3.

The fiber degeneration of the last-mentioned system is less massive than that observed in the amygdalo-hypothalamic group of fibers of Cat 1 (Fig. 4, section 3, An.len.). As shown in Fig. 5, section 3, these amygdalo-hypothalamic fibers rise upward from the dorsal pole of the area of coagulation. In Cat 1 I think the degeneration of the amygdalo-hypothalamic pathway conveys fibers originating outside the amygdaloid complex, too.

(c) From the dorsal pole of the lesion a moderate number of degenerating axons extend laterally to the gyrus sylvianus posterior (Fig. 5, section 3, G.S.P.). These fibers traverse the dorsal part of the amygdala, pierce the capsula externa (C.en.), and enter the white matter of the temporal cortex, forming small fascicles of coarse as well as fine degenerating elements, which can be followed to the tip of the temporal lobe. In the latter a moderate number of terminal fibers spread out diffusely. It should be mentioned that this amygdalo-temporal connection appears to be of minimal volume in the cat. No terminal degeneration was found in any other parts of the temporal lobe and the amygdalo-temporal fibers traced in this experiment were restricted exclusively to those parts of the temporal lobe identified as the homologue of the temporal tip in man and monkey.

(d) The degeneration observed in the present experiment coursing through the fasciculus longitudinalis associationis (Fig. 5, section 2, F.l.a.) is similar to that described in Cat 1. The condensed group of coarse degenerating fibers seen in the limit between regio praeoptica (R.P.O.) and area amygdaloidea anterior (A.a.) spreads dorsomedially, distributing numerous terminal fibers to the dorsolateral part of the regio praeoptica. As mentioned previously, it should be pointed out that in this region the area covered by the terminal degeneration of the fasciculus longitudinalis associationis overlaps partly with that supplied by the stria terminalis.

Scanty terminal degenerating fibers were observed in the area amygdaloidea anterior (Fig. 5, section 2, A.a.), lateral to the fasciculus longitudinalis associationis. These fibers form the diffuse part of the rostral projection system of the amygdala of which the fasciculus longitudinalis associationis forms its condensed or medial part. Further rostrally (section 1), degenerating fibers could be traced dorsal to the tuberculum olfactorium (Tu.Olf.), where fine degenerating terminals were observed. None appear to enter the cluster of cells of the tubercle. The tuberculum olfactorium is flanked medially by the diagonal band of Broca (D.B.B.), to which the rostral extension of the fasciculus longitudinalis associationis is joined. Moderate preterminal fibers and scanty coarse ones follow the diagonal band a short distance. This system of degenerating fibers, very similar to the rostral extension of the fasciculus longitudinalis associationis traced in Cat 1, appears, however, to end mainly in a group of cells located in the medial limits of the tuberculum olfactorium. This group of cells was identified as the nucleus of the diagonal band. No degeneration was traced to the regio limbica anterior as observed for Cat 1.

(e) Short intra-amygdaloid degenerating axons were traced from the lesion to the remainder of the amygdaloid nuclei (Fig. 5, section 3), excluding the nucleus amygdaloideus corticalis (A.co.). No degenerating fibers could be traced to the peri-amygdaloid cortex (Pir.). The bulk of the intra-amygdaloid connections observed in this experiment distribute themselves, however, to the nucleus amygdaloideus lateralis (A.l.), probably as degenerating collaterals of the amygdalo-temporal system of fibers.

CAT 4

In this experiment the lesion is situated in the ventral region of the amygdala. It encroaches slightly upon the underlying cortex piriformis (Pir.). As shown in Fig. 3, sections 8 to 10, and Fig. 6, section 3, the lateral part of the middle third of the rostral half of the nucleus amygdaloideus basalis, pars parvocellularis (A.b.p.), the adjoining parts of the same nucleus in its pars magnocellularis (A.b.m.), and the nucleus amygdaloideus lateralis (A.l.) were involved in the area of coagulation. The "frontal" sections of this brain are cut obliquely, the angle being dorso-

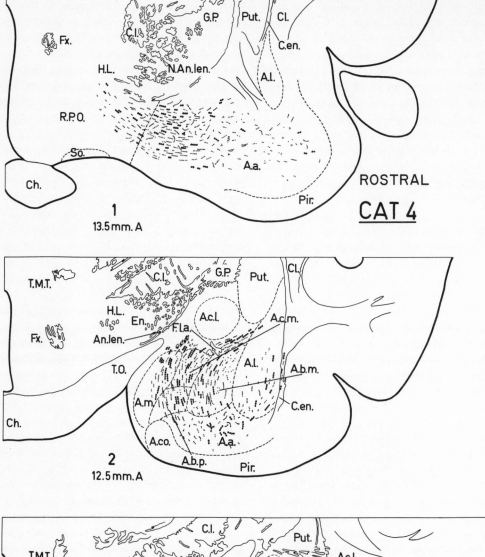

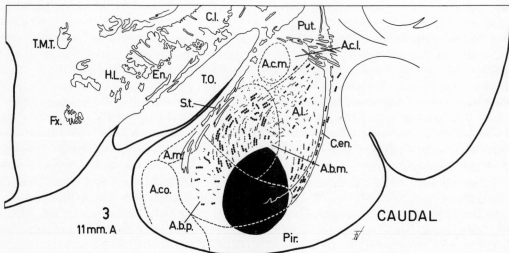

Fig. 6. Cat 4. Degeneration following lesion of the amygdaloid complex and underlying cortex piriformis.

caudally 30°–40° off from the frontal sections of the cat-brain atlases of Jasper and Ajmone-Marsan (1954) and Snider and Niemer (1961).

From the area of coagulation large numbers of degenerating fibers stream dorsally and rostrally, forming diffuse systems which can be divided into at least three groups: (a) rostrally, degenerating elements can be traced to the area amygdaloidea anterior and regio praeoptica; (b) dorsally, several condensed groups spread out in the dorsorostral regions of the amygdala; (c) a third system penetrates the capsula externa. In this experiment the lesion did not produce degeneration in the stria terminalis.

(a) The fasciculus longitudinalis associationis forms the condensed and medial part of a more diffuse and widespread fiber system: the rostral projection system of the amygdala, which extends from the amygdaloid complex rostrally to the area amygdaloidea anterior and the preoptic region and farther rostrally to the diagonal band and the substantia perforata just dorsal to the tuberculum olfactorium. Still farther rostrally it reaches the regio limbica inferior. As was observed in Cats 1 and 3, all these structures are supplied by the condensed part of the rostral projection system of the amygdala, namely, the fasciculus longitudinalis associationis. However, besides this condensed group of fibers, diffuse systems were traced in Cats 1 and 3 to the area amygdaloidea anterior and cortex praepiriformis. The degenerating elements observed in the regio praeoptica (R.P.O.) and area amygdaloidea anterior (A.a.) in Cat 4 (Fig. 6, section 1) represent a very complete degeneration picture of the lateral or diffuse part of the rostral projection system of the amygdala. The fasciculus longitudinalis associationis is free from degenerating elements (see Fig. 6, section 2, F.l.a.).

(b) A second system of degenerating fibers was traced from the dorsal aspect of the lesion (section 3) distributing numerous terminal fibers to the amygdaloid nuclei: medialis (A.m.), pars medialis of the centralis (A.c.m.), basalis (A.b.m. and A.b.p.), and lateralis (A.l.). The appearance of this system of degenerating fibers resembles closely the fantail system observed in Heidenhain-stained sections; compare Fig. 6, sections 2 and 3, with Fig. 47, system 1 or capsula intermedia (C.im.).

In Nauta sections (Fig. 6, section 2) this system has the appearance of small bundles of coarse, as well as fine, preterminal fibers that ascend dorsally to the dorsal limits of the amygdala, where these small fascicles curve laterally toward the putamen (Put.) between the pars lateralis of the central nucleus (A.c.l.) and the medial border of the nucleus amygdaloideus lateralis (A.l.).

The degenerating fibers of this fantail system pierce the pars magnocellularis of the basal nucleus (A.b.m.) and, rostrally, the dorsolateral part of the nucleus amygdaloideus medialis (section 2, A.m.), the pars medialis of the nucleus amygdaloideus centralis (A.c.m.), and the dorsal zone of the parvocellular division of the basalis (A.b.p.). It was observed that part of the fiber degeneration of this fantail system ends in the rostral part of the nucleus amygdaloideus lateralis (section 3, A.l.). My Golgi observations (see Figs. 45 and 46) suggest that this system originates mainly in the periamygdaloid cortex.

(c) From the lateral aspect of the lesion a moderate number of coarse, as well as fine, degenerating fibers course in a dorsal direction, the former through the capsula externa (Fig. 6, sections 2 and 3, C.en.), the latter as a small group of fibers through the ventrolateral part of the nucleus amygdaloideus lateralis (sections 2 and 3, A.l.). Degeneration in the capsula externa could not be followed farther. These fibers represent degenerating axons of cells of the cortex piriformis that penetrate the amygdala through the capsula externa on its lateral side. Hence, piriform-amygdaloid connections are established in part through the capsula externa. Further evidence of these connections is presented in Golgi material (see pp. 86–92 and Figs. 45, 46, and 48).

CAT 5

The brain of this animal was studied in sagittal sections. As shown by Fig. 3, sections 5 to 8, and Fig. 7, sections 2 to 4, the area of coagulation comprises the caudoventral part of the area amygdaloidea anterior (A.a.) and the rostral third of the nucleus amygdaloideus basalis (A.b.m. and A.b.p.), with slight involvement of the adjoining medioventral part of the nucleus amygdaloideus lateralis (A.l.). Encroachment upon the underlying

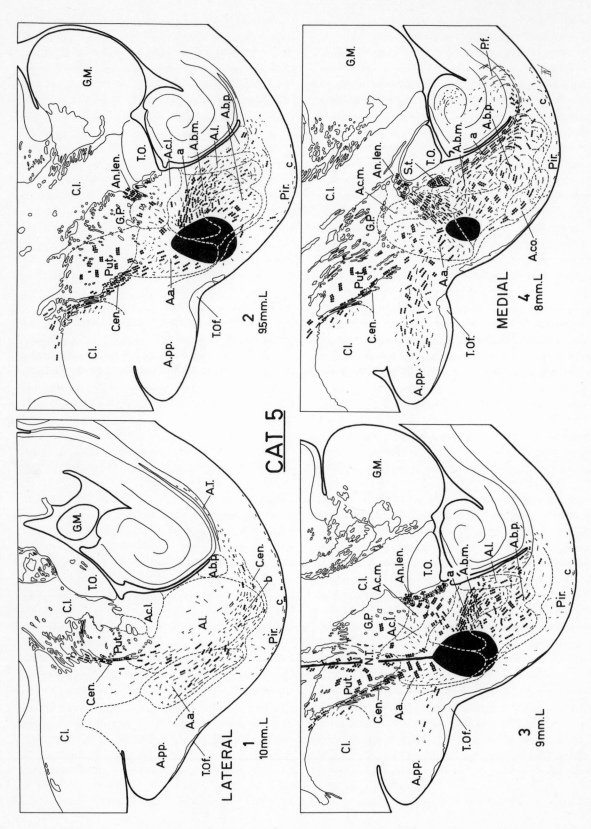

Fig. 7. Cat 5. Degeneration following lesion of the amygdaloid complex and underlying cortex piriformis. Sagittal sections.

cortex piriformis (Pir.) also was observed. The needle tract interrupted some fibers of the capsula interna (C.I.) and touched part of the putamen (Put.).

Six projection systems were observed: (a) rostrally to the area amygdaloidea anterior and cortex praepiriformis through the lateral, diffuse component of the rostral projection system of the amygdala; (b) rostrodorsally through the capsula externa to the claustrum; (c) rostromedially to the regio praeoptica and diagonal band by the route of the medial, compact component of the rostral projection system of the amygdala, namely, the fasciculus longitudinalis associationis; (d) to the regio praeoptica, the base of the septum, the nucleus accumbens septi, and the hypothalamus throughout the stria terminalis; (e) to the caudal part of the nucleus medialis dorsalis of the thalamus by way of the inferior thalamic peduncle; finally (f) a fantail system of intra-amygdaloid fibers shows massive fiber degeneration extending caudoventrally to the remainder of the amygdaloid nuclei, cortex piriformis, and hippocampus.

(a) In Figs. 7 and 8 degenerating fibers can be seen extending rostromedially through the lateral or diffuse component of the rostral projection system of the amygdala. In sections 1 to 5 the area amygdaloidea anterior (A.a.) is pervaded by numerous degenerating fine axons. In sections 3 and 4 (Fig. 7) a rather condensed group of coarse as well as fine fibers can be traced to the area praepiriformis (A.pp.). In the latter the degeneration does not reach the superficial layer, but it is circumscribed in the depth of this region. The group of coarse degenerating fibers observed in Fig. 8, section 5, between the nucleus amygdaloideus medialis (A.m.) and the area amygdaloidea anterior (A.a.), which also distributes fine fibers to the nucleus tractus olfactorii lateralis (N.T.Of.), represents fibers of the lateral part of this rostral projection system of the amygdala.

(b) Coarse fibers can be traced in sections 1 to 4 running in the capsula externa (Fig. 7, C.en.). In section 3, this projection courses rostrodorsally through the area amygdaloidea anterior, entering the capsula. In sections 2 to 4 some fine degenerating fibers can be seen ending in the caudal pole of the claustrum (Cl.). Coarse degenerating elements were observed in the putamen (Put.), but they are probably fibers interrupted by the needle tract. Further rostrally, coarse fibers, which have been traced through the capsula externa, enter the white matter of the brain (Fig. 7, sections 2 to 4) and could not be followed further. These fibers would represent probably a connection between the amygdala and the gyrus orbitalis. Its course through the capsula externa is quite similar to that followed in the opposite direction by the orbito-amygdaloid fibers of Cat 19; see paragraph (b) and Figs. 22 and 23. However, the definitive demonstration of a reciprocal connection between the gyrus orbitalis and the amygdala remains opened for further investigations.

(c) As shown in Figs. 8 and 9 (sections 5 to 9), coarse fibers collect in condensed groups just ventral to the commissura anterior (C.A.), forming the fasciculus longitudinalis associationis (F.l.a.), which distributes terminal fibers underneath the globus pallidus (section 5, G.P.), as well as to the caudal part of the tuberculum olfactorium (sections 5 and 6, Tu.Olf.). More medially the regio praeoptica (sections 6 and 7, R.P.O.) is pervaded by a moderate number of fine degenerating fibers which partly overlap with the degeneration traced still more medially (section 8) to the base of the septum (Spt.) via the stria terminalis (S.t.). Finally, the coarse degenerating fibers of the fasciculus longitudinalis associationis (F.l.a.), just dorsal to the diagonal band (D.B.B.) in section 8, were traced to the vertical limb of this band (section 9) and more rostrally to the white matter of the gyrus rectus and regio limbica anterior.

(d) The stria terminalis (Fig. 7, section 4, S.t.) is degenerated in its intra-amygdaloid course. Part of the coarse degenerating fibers, extending from the area of coagulation toward the tractus opticus (T.O.) in section 3, collect in the compact bundle of the stria terminalis (S.t.) in section 4, showing clearly how part of the stria terminalis originates in those areas of the nucleus amygdaloideus basalis (A.b.m. and A.b.p.) involved in the lesion in this experiment. At commissural levels two components of the stria terminalis were observed and identified as the S.t.3 and S.t.4 components of Johnston (1923).

The precommissural or S.t.4 component dis-

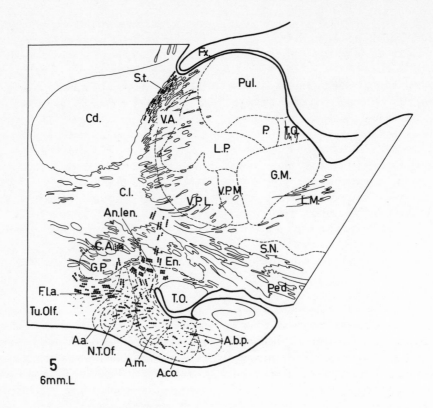

LATERAL

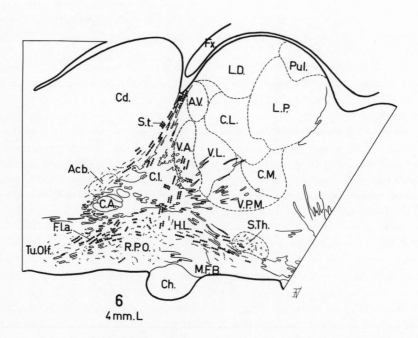

MEDIAL

Fig. 8. Cat 5. Degeneration following lesion of the amygdaloid complex and underlying cortex piriformis, as indicated by Fig. 7. Sagittal sections.

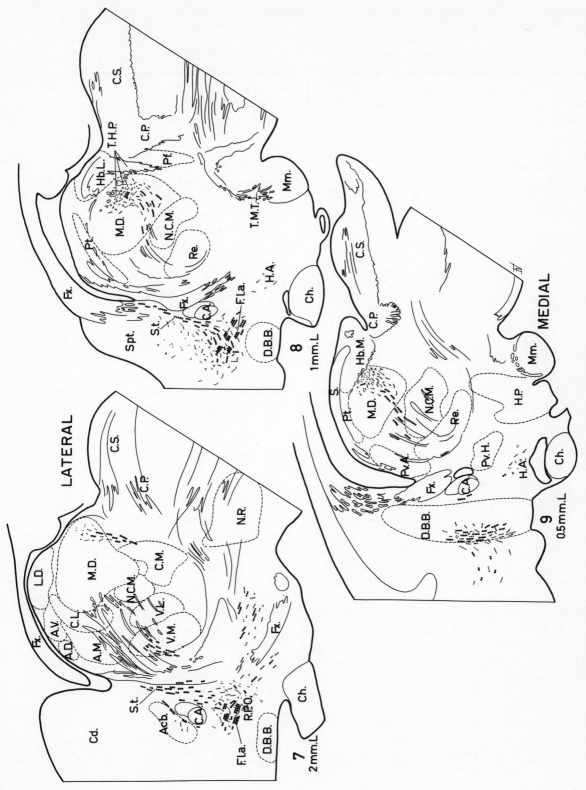

LATERAL

Cd.

C.S.

C.P.

L.D.

M.D.

Fx.

A.V.

A.D.

C.L.

N.C.M.

C.M.

A.M.

V.L.

V.M.

N.R.

Fx.

S.t.

Acb.

C.A.

Ch.

F.la.

R.P.O.

D.B.B.

7
2mm.L

C.S.

T.H.P.

Hb.L.

C.P.

Pf.

Fx.

Pt.

M.D.

N.C.M.

Mm.

Re.

T.M.T.

Spt.

Fx.

H.A.

C.A.

F.la.

Ch.

S.t.

D.B.B.

8
1mm.L

C.S.

C.P.

S

Pt.

Hb.M.

M.D.

N.C.M.

Re.

H.P.

Mm.

Pv.A.

Fx.

Pv.H.

C.A.

H.A.

Ch.

D.B.B.

MEDIAL

9
0.5mm.L

Fig. 9. Cat 5. Degeneration following lesion of the amygdaloid complex and underlying cortex piriformis, as indicated by Fig. 7. Sagittal sections.

21

tributes a moderate number of terminal fibers to the nucleus accumbens septi (Figs. 8 and 9, sections 6 and 7, Acb.) and to the base of the septum (section 8, Spt.) rostral to the commissura anterior (C.A.). The terminal fibers of this component, as mentioned before, partly overlap in the dorsal part of the regio praeoptica with the endings of the fasciculus longitudinalis associationis.

The postcommissural or S.t.3 component distributes fibers to the regio praeoptica (section 7, R.P.O.), where, like the S.t.4 component, it also overlaps with the preoptic termination of some of the fibers traced by way of the rostral projection system of the amygdala.

More caudoventrally, fibers coursing through the medial forebrain bundle (Fig. 8, section 6, M.F.B.) are present in the hypothalamus lateralis (Fig. 8, section 6, H.L.) and in the nucleus subthalamicus (Fig. 8, section 6, S.Th.). More medially (Fig. 9, section 7), degenerating fibers were also observed in the adjoining dorsal and ventral regions of the hypothalamic fornix (Fx.). Scanty, fine degenerating fibers were observed still more medially in sections 8 and 9 in the hypothalamus anterior (H.A.); it was difficult, however, to decide if these fibers represent the hypothalamic or S.t.2 component of the stria terminalis, or if they come from the rostral projection system of the amygdala, or even if they form part of a direct amygdalo-hypothalamic connection, unrecognized in this sagittal series.

(e) Coarse degenerating fibers (Fig. 7, sections 2 to 4) extend from the lesion dorsally to the zone just rostral to the tractus opticus (T.O.), where they join the ansa lenticularis (An.len.). More medially (Fig. 8, section 5), these elements pierce the nucleus entopeduncularis (En.) and, going upward through the capsula interna (C.I.), enter the nucleus ventralis anterior of the thalamus (section 6, V.A.), without showing signs of termination in this structure, and join the inferior thalamic peduncle at the level of the nucleus ventralis medialis (Fig. 9, section 7, V.M.). In this section a moderate number of coarse fibers enter the nucleus medialis dorsalis (M.D.) and distribute scanty terminal elements in the caudal part of this thalamic nucleus. As shown by sections 7, 8, and 9, this sytem of fibers follows an S-like course in its rostrocaudal projection through the thalamus. The system ends in the posteromedial part of the nucleus medialis dorsalis just rostral to the tractus habenulo-peduncularis (T.H.P.).

It should be noted that the number of degenerating elements observed in the nucleus medialis dorsalis is far smaller than that in the massive fiber degeneration detectable in the amygdala, extending from the area of coagulation rostrally to the tractus opticus where this system joins the ansa lenticularis (see Fig. 7, section 4). Undoubtedly, other fibers, which have escaped recognition, are distributed to the hypothalamus through the sublenticular projection of the ansa lenticularis before the latter joins the inferior thalamic peduncle.

It seems likely that the neurones which give rise to this amygdalo-thalamic connection are located in the area amygdaloidea anterior. Little, if any, contribution can be expected from the remaining amygdaloid nuclei. Cats with lesions in the amygdaloid nuclei sparing the ventral part of the area amygdaloidea anterior never show fiber degeneration traceable to the nucleus medialis dorsalis of the thalamus.

(f) A considerable number of coarse, as well as fine, degenerating fibers spread out in a caudoventral direction from the lesion throughout the remainder of the amygdaloid nuclei, the underlying cortex piriformis, and further caudalward to the hippocampus. This degeneration, which in part follows a fantail intra-amygdaloid course, could be grouped into at least three systems which were labeled a, b, and c in Fig. 7, sections 1 to 4.

System a, formed by densely packed coarse and fine fibers, projects caudally from the dorsal pole of the area of coagulation toward the floor of the lateral ventricle, where it continues as a thin sheet of degenerating fibers to the ventricular edge in the caudal limits of the amygdala (section 4). This sytem, before reaching the floor of the ventricle, distributes fine terminal degenerating fibers, which spread out over the pars magnocellularis of the basal (A.b.m.) and caudal part of the lateral (A.l.) amygdaloid nuclei (sections

2 and 3). More caudally, where this system approaches the floor of the ventricle (section 4), it distributes terminal fibers to the caudal part of both pars magnocellularis and parvocellularis of the nucleus amygdaloideus basalis (A.b.m. and A.b.p.). In the caudal ventricular border the few remaining fibers of the system enter the angular tract of the hippocampus, turn dorsally (section 4, P.f.), pierce the stratum moleculare, and enter the CA4 field of Ammon's horn, where they terminate.

System *b* (Fig. 7, section 1) follows a more ventral course, occupying initially the ventral part of the capsula externa (C.en.), where a moderate number of fine fibers leave to enter the ventral region of the nucleus amygdaloideus lateralis (A.l.). More caudally a few fine degenerating fibers enter the angular tract (A.T.), where they could not be followed farther.

Just under system *a* there are other fibers (not labeled) which form diffuse and isolated little fascicles of coarse, as well as fine, degenerating elements, terminating caudally in the pars parvocellularis of the nucleus amygdaloideus basalis (sections 2 to 4, A.b.p.).

System *c* (Fig. 7, sections 1 to 4) takes a ventral exit from the lesion, traverses the entire thickness of the cortex piriformis (Pir.), and reaches the superficial fibrillar layer, where a few fine degenerating fibers could be traced to the entorhinal cortex. Coarse fibers of system *c* pierce the nucleus amygdaloideus corticalis (section 4, A.co.), where a few fibers terminate. This system probably does not originate in any part of the amygdala, but instead probably represents recurrent collaterals of axons of pyramidal cells as well as ascending axons of the stellate and polymorphic cells of the cortex piriformis, which contribute to form the fibrillar layer of this cortex.

CAT 6

The lesion (Fig. 3, sections 11 to 16, and Figs. 10 and 11, sections 3 to 6) is the largest in this experimental series. It involves a considerable portion of the amygdala and of the tractus opticus (T.O.), and caudally it encroaches upon the hippocampal formation. In the amygdala (Figs. 10 and 11, sections 3 to 5) the lesion destroys part of the nucleus amygdaloideus centralis, pars medialis and pars lateralis (A.c.m. and A.c.l.), as well as part of the dorsal and medial zones of the basal, pars magnocellularis (A.b.m.), and lateral (A.l.) amygdaloid nuclei. The lesion also involves (Fig. 10, section 3) the compact part of the intra-amygdaloid course of the stria terminalis (S.t.) and the fasciculus longitudinalis associationis (F.l.a.). The degeneration caused by the involvement of the tractus opticus and the hippocampus (Fig. 11) will not be considered here.

Five systems of fibers could be traced from the amygdaloid lesion: (*a*) the stria terminalis; (*b*) the rostral projection system of the amygdala through both its medial, compact part, or fasciculus longitudinalis associationis, and its diffuse, or lateral, part; (*c*) fibers to the hypothalamus lateralis by way of the medial amygdalo-hypothalamic pathway; (*d*) lateral fibers to the gyrus sylvianus posterior; and finally (*e*) a few fibers to the remainder of the amygdaloid nuclei, excluding the nucleus amygdaloideus corticalis.

(*a*) In spite of the great size of the lesion, the degeneration in the stria terminalis is rather moderate. As it approaches the commissura anterior, the stria terminalis spreads out, distributing fine degenerating fibers to the bed nucleus of the stria (Fig. 10, section 1, B.S.t.). Other degenerating fibers, traced in this experiment via the stria terminalis, enter the commissura anterior (section 1, C.A.), where a moderate number of fine and coarse fibers could be followed around the outermost stratum of the commissura. As will be pointed out later, these fibers establish synaptic contact with dendrites protruding into the commissura from cells of the surrounding bed nuclei of the stria terminalis and commissura anterior. No stria degeneration could be traced ventrally to preoptic or hypothalamic levels.

(*b*) The fasciculus longitudinalis associationis (Fig. 10, section 2, F.l.a.) shows abundant degenerating coarse fibers, which could be followed to the area amygdaloidea anterior (Fig. 10, section 1, A.a.) and regio praeoptica (section 1, R.P.O.), especially its lateral part. The medial part of the area amygdaloidea anterior, supplied by terminal fibers of the fasciculus longitudinalis associationis, is also supplied with scattered coarse

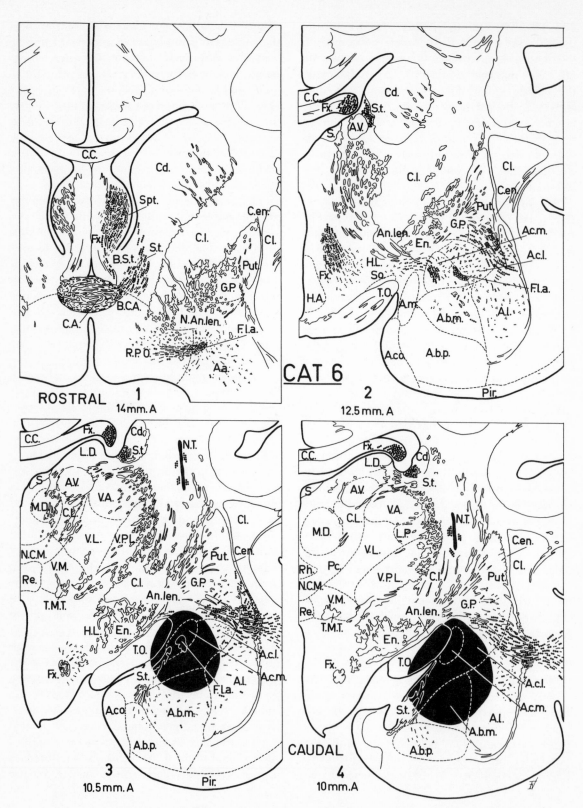

Fig. 10. Cat 6. Degeneration following lesion of the amygdaloid complex, tractus opticus, and ventral hippocampus, as indicated in this figure and Fig. 11.

and fine fibers traced diffusely from the amygdala. As was mentioned in the preceding experiments (see Cats 1 and 3), this latter system connects the amygdaloid complex with the area praepiriformis and amygdaloidea anterior by a system of short-axoned neurons that project rostrally. When the lesion is placed more rostrally, as occurs in Cat 5, these short links could be traced farther rostrally to the area praepiriformis. Thus, it appears that this lateral, diffuse component of the rostral projection system of the amygdala is made up exclusively of short axons.

(c) In the medial zone of the nucleus amygdaloideus centralis, pars medialis (Fig. 10, section 2, A.c.m.), a group of coarse degenerating fibers collect and thence, following in part the ansa lenticularis, enter the hypothalamus lateralis (H.L.), where scanty terminal degeneration was observed just dorsal to the nucleus supraopticus (So.). A few fibers of this system can be observed coursing in the ansa lenticularis (section 3, An.len.). This degeneration appears to be similar to that described in Cats 1 and 3 as forming the medial amygdalo-hypothalamic pathway.

(d) From the dorsal pole of the area of coagulation a massive system of coarse fibers (Fig. 10, sections 2, 3, and 4) extends laterally along the dorsal limits of the amygdala. These fibers traverse the base of the putamen (Put.), pierce the dorsalmost tip of the nucleus amygdaloideus lateralis (sections 3 and 4, A.l.), and project laterally through a zone in the capsula externa (C.en.) that appears fringed in normal Heidenhain-stained sections. This part of the capsula externa, made up of interlacing fibers coursing in various directions, probably represents the hilus of the gyrus sylvianus posterior. From here the degeneration spreads out in the white matter of the gyrus, and terminal degeneration (not represented) was found in the deepest cortical layers. The question arises whether or not the lesion has interrupted extra-amygdaloid corticopetal fibers entering the gyrus sylvianus posterior through a sublenticular transit. This is possible, but some of these fibers represent amygdalo-cortical connections, as shown previously in the experiment on Cat 3 in which these connections were traced directly from the lesion to the temporal cortex without involvement of any other corticopetal fibers originating outside the amygdala. The degeneration in the temporal lobe ends exclusively in those parts that correspond comparatively to the temporal pole of primates and man.

(e) Numerous fibers of various calibers are distributed throughout the amygdaloid nuclei and probably represent degenerated intra-amygdaloid collaterals of axons of several types of cells which interconnect diffusely different amygdaloid areas. These fibers are particularly abundant in both subdivisions of the nucleus amygdaloideus centralis (A.c.l. and A.c.m.). They can also be observed, although not so abundantly, in parts of the basal and lateral amygdaloid nuclei; however, in these nuclei there are zones free of degenerating elements. No degenerating fibers were observed entering the nucleus amygdaloideus corticalis (A.co.).

The recognition that there is a considerable zone of the basolateral amygdala entirely free from degenerating fibers relates nicely with my Golgi material suggesting a preferential dorsomedial direction of the axons of cells of the basolateral amygdaloid areas (see Figs. 46 and 48). Golgi observations also reveal that the nucleus amygdaloideus corticalis is a way station between the axons of olfactory mitral cells and the stria amygdaloid cells. Few, if any, axons of cells of the medial and central nuclei of the amygdala reach the nucleus amygdaloideus corticalis.

CAT 8

In this experiment the brain was cut horizontally. An attempt was made to destroy the tractus olfactorius lateralis in order to trace its contributions to the amygdala. As shown in Fig. 3, sections 3 and 4, and in Fig. 12, sections 2 and 3, this was only partly accomplished, as a considerable portion of the tractus olfactorius was spared. The lesion penetrated into the cortex piriformis and reached slightly the area amygdaloidea anterior.

Two systems of fibers were observed: (a) a superficial and (b) a deep group of fibers.

(a) The superficial stream forms part of the fibrillar layer of the cortex piriformis and can be clearly seen in section 1 of Fig. 12 (F.L.); it is

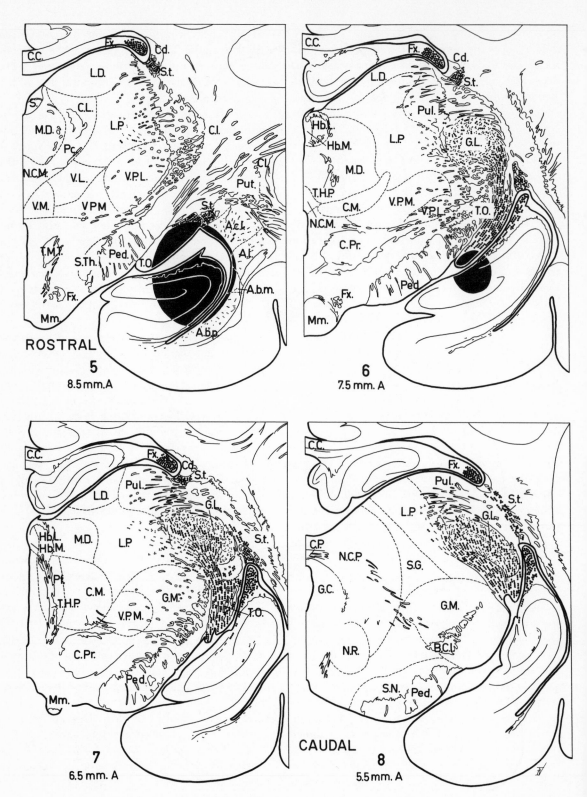

Fig. 11. Cat 6. Degeneration following lesion of the amygdaloid complex, tractus opticus, and ventral hippocampus, as indicated in this figure and Fig. 10.

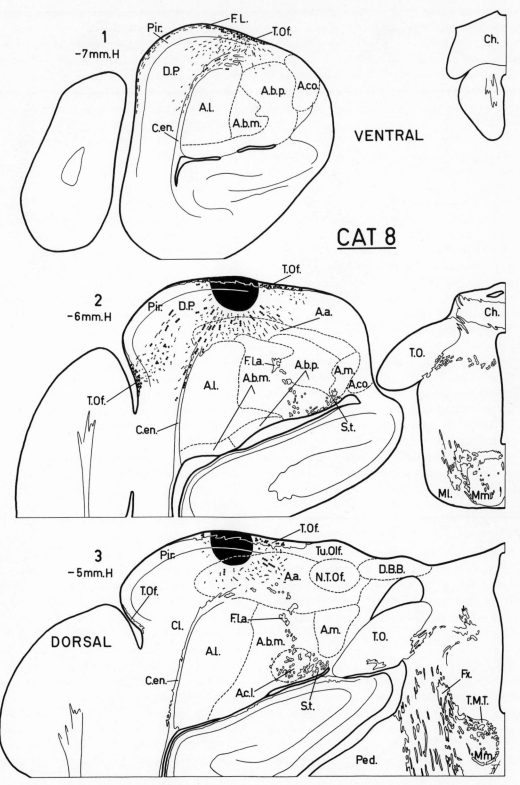

Fig. 12. Cat 8. Degeneration following lesion of the tractus olfactorius lateralis and overlying cortex piriformis. Horizontal sections.

composed of a moderate number of fine degenerating fibers and a scarce group of coarse ones. This system of fibers does not extend through the entire cortex piriformis. At the level of the middle of the amygdala (section 2) it turns rostrally, entering the deep plexus (D.P.) of the cortex piriformis between the latter (Pir.) and the capsula externa (C.en.).

(b) The deep system of fibers courses directly caudally from the lesion, spreading out in fantail fashion and reaching a considerable extension of the lateral half of the area amygdaloidea anterior (sections 2 and 3, A.a.). Part of these fibers enter the rostralmost zone of the capsula externa (section 1, C.en.) where the latter fuses medially with the deep plexus of the cortex piriformis. Fine, as well as some coarse, degenerating fibers course caudally through this capsule and appear to end in the adjoining deep zones of the cortex piriformis (sections 1 and 2).

No degeneration was traced to any part of the amygdaloid nuclei, or to the nucleus tractus olfactorii lateralis (N.T.Of.), the tuberculum olfactorium (Tu.Olf.), or the diagonal band (D.B.B.). These centers, as well as the nucleus amygdaloideus corticalis (A.co.), are supplied by more medial fibers of the tractus olfactorius lateralis than those involved in the lesion of this experiment.

The results obtained in the present experiment show that the amygdala, excluding the area amygdaloidea anterior, is supplied indirectly by the tractus olfactorius through the cortex piriformis.

CAT 11

In this animal an attempt was made to place a lesion in the course of fibers traversing both medial and lateral parts of the regio praeoptica. After the passage of the current for a shorter time than in other experimental animals, the electrode was raised 4 mm and the current was again passed for a short time.

As shown by Fig. 3, section 3, and Fig. 13, section 2, the area of coagulation interrupted many fibers ascending from the regio praeoptica to the base of the septum and vice versa. The lesion encroached upon the commissura anterior where the temporal and bulbar components of the latter join each other. The medial part of the nucleus accumbens septi (Acb.) and adjoining parts of the septum also were damaged.

Four systems of fibers could be traced from the lesion: (a) fibers to and from the base of the septum and the diagonal band of Broca; (b) commissural fibers through both temporal and bulbar components; (c) a small projection through the stria medullaris; and finally (d) a system of fibers that reach the hippocampal formation via the fornix.

(a) As can be seen in Fig. 13, section 2, massive fiber degeneration extends dorsomedially to the base of the septum (Spt.) and the medial limb of the diagonal band (D.B.B.), the former composed of degenerating fibers of heavy caliber, the latter of dustlike fine disintegrating elements. In both systems the alignment of the fibers is remarkable. The coarse lateral system of fibers appears to end by supplying a moderate number of fibrils in a restricted zone of the base of the septum just rostral to the fornix and a little above the commissura anterior.

The medial system of fine fibers can be identified within the diagonal band of Broca. This latter, made up of numerous short connections, forms a continuous band of short axons and cells extending from rostromedial amygdaloid regions throughout the base of the regio praeoptica, where the diagonal band curves and ascends through the medial wall of the hemisphere. No long axons course in this band and, thus, the fine degenerating fibers that form this medial system represent the disintegrating axons of cells of that part of the diagonal band destroyed by the lesion.

At levels caudal to the lesion (Fig. 13, section 3), numerous fine and some coarse fibers enter the regio praeoptica (R.P.O.). Part of these fibers represent short connections of the diagonal band coursing toward the amygdala. In section 3 the degeneration in the regio praeoptica (R.P.O.) forms a complicated maze and numerous fine terminal fibers appear to synapse here. Some of them penetrate the nucleus supraopticus (So.), where they terminate.

From these observations it is clear that the fibers of heavier caliber which, medial to the lesion, ascend to the base of the septum represent the ascending part of the Zuckerkandl's radiation.

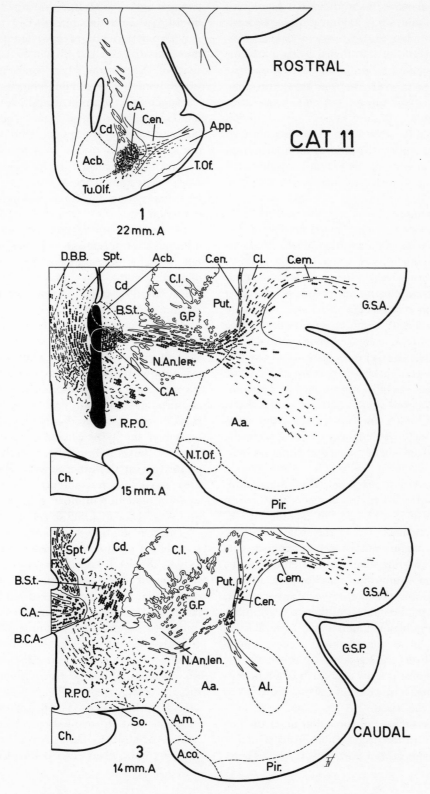

Fig. 13. Cat 11. Degeneration following lesion of the regio praeoptica.

The two systems of fibers, diagonal band and Zuckerkandl's radiation, both containing ascending and descending fibers, course in close association with each other and medial to the pars bulbaris of the commissura anterior, even though at preoptic levels they diverge from each other, the former toward the amygdaloid region, the latter as one of the major contributions of the medial forebrain bundle area. The ascending part of both systems could be considered an indirect and rostral continuation of the rostral projection system of the amygdala, which relays in the regio praeoptica.

The descending fibers of the Zuckerkandl's radiation, which are concentrated laterally to the lesion (section 2) in three little fascicles of coarse fibers, spread out at more caudal levels (section 3), where numerous wavy, fine degenerating fibers and some coarse ones occupy a considerable part of the regio praeoptica (R.P.O.). Farther caudally (Fig. 14, section 4), a moderate number of fine fibers are still present in the hypothalamus lateralis (H.L.). The degeneration present in these two regions, namely, the regio praeoptica and the hypothalamus lateralis, represents the contribution of the Zuckerkandl's radiation to the short-axoned cells which, throughout the regio praeoptica and the hypothalamus lateralis, form the so-called bed nucleus of the medial forebrain bundle.

It is evident that the lesion in this animal damaged other diverse systems of fibers that course through this region. I could not exclude a probable involvement of part of the stria terminalis, especially in its precommissural component. The small group of fine fibers seen lateral and just ventral to the commissura anterior (Fig. 13, section 2, C.A.) and the fine degenerating elements observed in the bed nuclei of the stria terminalis and of the commissura anterior (B.S.t. and B.C.A., respectively, in section 3) might represent such fibers. Fine degenerating fibers are present in the nucleus accumbens septi (Acb.); they might represent collaterals of axons of cells of this nucleus involved in the lesion.

The encroachment of the lesion upon the fornix gives rise to the degeneration in its postcommissural course (Fig. 14, section 4, Fx.). There-fore, part of the fine degenerating fibers observed dorsal to the tractus opticus (T.O.), between the hypothalamus anterior (H.A.) and the hypothalamus lateralis (H.L.), represent the termination of some of these fornix fibers which in part interweave with the descending contributions of the Zuckerkandl's radiation to the medial forebrain bundle area. No fornical degeneration was traced farther caudally. It may be possible that some descending fibers of the stria terminalis, which could not be identified in this animal, are actually present among the terminal degenerating fibers observed in the regio praeoptica caudal to the lesion.

(b) As shown in Fig. 13, section 2, the lesion involved a considerable portion of the commissura anterior where the pars bulbaris and temporalis join each other. This caused damage of fibers which could be traced in both commissural segments.

The degeneration through the pars temporalis could be followed to the base of the putamen (section 2, Put.) where two main streams of degenerating fibers diverge from each other; one of them (sections 2 to 4) ascends through the outer capsules and the claustrum (Cl.), in which terminal degeneration is abundant, enters the white matter of the gyrus sylvianus anterior (G.S.A.), and distributes terminal fibers to the deep layers of this gyrus; the other stream descends to the area amygdaloidea anterior (section 2, A.a.), where a moderate number of terminal fibers end around cells of this diffuse cellular region.

The degeneration traced through the pars bulbaris of the commissura anterior (Fig. 13, section 1) was followed rostrally as a compact bundle made up of coarse fibers. Just rostral to the putamen and at the region where the capsula externa (C.en.) approaches the commissura anterior (C.A.), numerous degenerating terminal fibers leave the commissura and, coursing laterally through the capsula externa (C.en.), enter the deep plexus of the area praepiriformis (A.pp.), where they terminate. Scanty terminal fibers descend to the tuberculum olfactorium (Tu.Olf.). None were traced medially, that is, to the rostral extension of the nucleus accumbens septi (Acb.). Still more rostrally terminal fibers enter the nucleus olfactorius

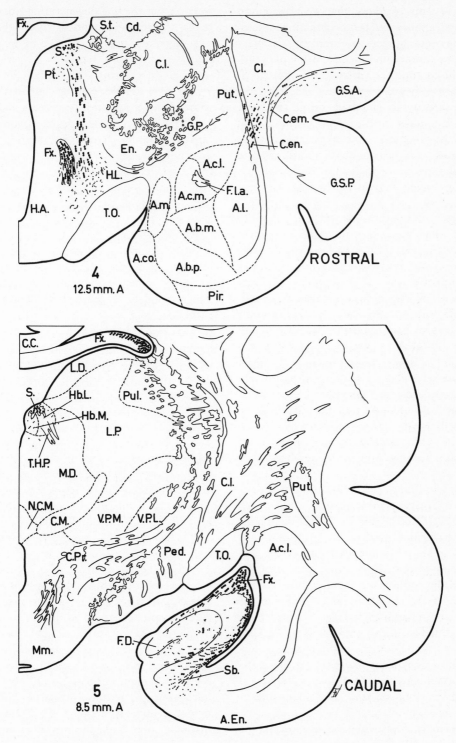

Fig. 14. Cat 11. Degeneration following lesion of the regio praeoptica, as indicated in Fig. 13.

anterior (not represented) and, finally, some fibers penetrate the olfactory bulb where they terminate among the internal granular cells (not represented).

The degeneration traced through both pars bulbaris and pars temporalis of the commissura anterior is bilateral. It was observed that contralaterally the pattern of distribution in the two segments of the commissura was the same as that just described. No predominance of degenerating fibers on either side was observed.

(c) Caudal to the lesion (Fig. 13, section 3) a group of coarse fibers was observed between the commissura anterior (C.A.) and the capsula interna (C.I.). This group of fibers could be identified as one of the components of the stria medullaris; these fibers (Fig. 14, section 4) ascends lateral to the fornix (Fx.), a small number terminate in the nucleus parataenialis (Pt.), and the rest, a rather scanty group of coarse fibers, continue in the stria medullaris (S.) and terminate in the rostral part of the habenular group of nuclei (section 5, Hb.L. and Hb.M.). Scanty terminal fibers were traced to the adjoining region of the nucleus medialis dorsalis (M.D.). The origin of this small component of the stria medullaris could not be definitely determined.

(d) The lesion damaged the fornix where it bends behind the commissura anterior (Fig. 13, section 3) and the adjoining septal region. This caused degeneration of coarse fibers which could be followed backward to the ventral hippocampal formation (Fig. 14, section 5), where the fibers enter the alveus and distribute a moderate number of terminal ones to the strata pyramidale and oriens, the dentate gyrus, and the subiculum.

CAT 15

In this animal, as shown by Fig. 3, sections 13 to 16, and Fig. 16, sections 7 and 8, the lesion was situated in the lamina medullaris medialis of the thalamus with involvement of the dorsal part of the intralaminar nucleus centralis lateralis (C.L.) and parts of the adjoining medialis dorsalis (M.D.), lateralis dorsalis (L.D.), and lateralis posterior (L.P.) thalamic nuclei. Additional damage to the gyri centralis and cinguli, the corpus callosum, and the fornix was caused by the nee-

dle tract; the cingulum (Fig. 16, section 8, Cng.) was also damaged.

Massive fiber degeneration extends from the lesion rostrally through the lamina medullaris medialis to rostral thalamic levels. At the level of the lesion (Fig. 16, section 8), two main streams of coarse degenerating fibers leave the area of coagulation. One of them projects in a dorsolateral direction toward the tail of the caudate nucleus (Cd.); the fibers enter the capsula interna (C.I.) just ventral to the caudate, and disperse in various directions. Along their intrathalamic course, these fibers provide scanty terminal degeneration to the thalamic nuclei lateralis dorsalis (L.D.) and lateralis posterior (L.P.).

The other system takes directly a rostral direction through the lamina medullaris medialis (Fig. 16, section 7, L.M.M.), where the degenerating fibers have been cut transversally. Fine and coarse degenerating elements spread out diffusely from the lesion and partly invade the medial zones of the nucleus lateralis posterior of the thalamus (section 7, L.P.), a considerable zone of the lateral part of the nucleus medialis dorsalis (sections 7 and 8, M.D.), and finally other, rather scanty, terminal fibrils enter the nucleus centrum medianum (section 8, C.M.) and nucleus ventralis lateralis (section 7, V.L.). At more rostral thalamic levels (section 6) the lamina medullaris medialis (L.M.M.) bifurcates, partly surrounding the anterior part of the thalamus; medially, the fibers enter the lamina medullaris superficialis (L.M.S.); laterally, the fibers approach the ventral surface of the nucleus caudatus (Cd.). At the point of bifurcation massive coarse degenerating fibers are present. Some of these fibers travel through the medial parts of the nucleus ventralis anterior (V.A.) and the dorsomedial part of the nucleus ventralis lateralis (V.L.). In these two nuclei terminal degenerating fibers abound.

Fine and coarse fibers project through the lamina medullaris superficialis (Fig. 16, sections 5 and 6, L.M.S.) and a moderate number enter the nucleus lateralis dorsalis (L.D.) through its superficial stratum; other fibers appear to continue more laterally to join the pedunculus thalami dorsalis. The latter, extending between the point of bifurcation of the lamina medullaris medialis

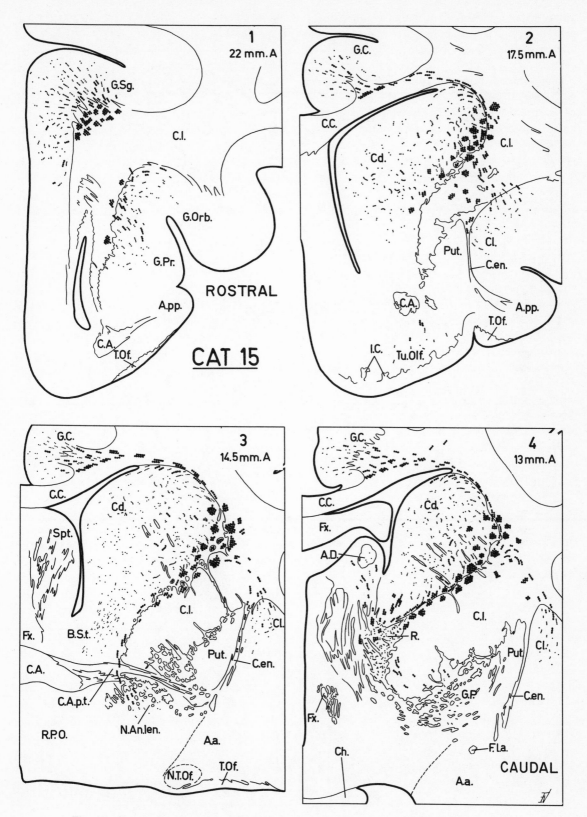

Fig. 15. Cat 15. Degeneration following lesion of the thalamus as indicated in Fig. 16.

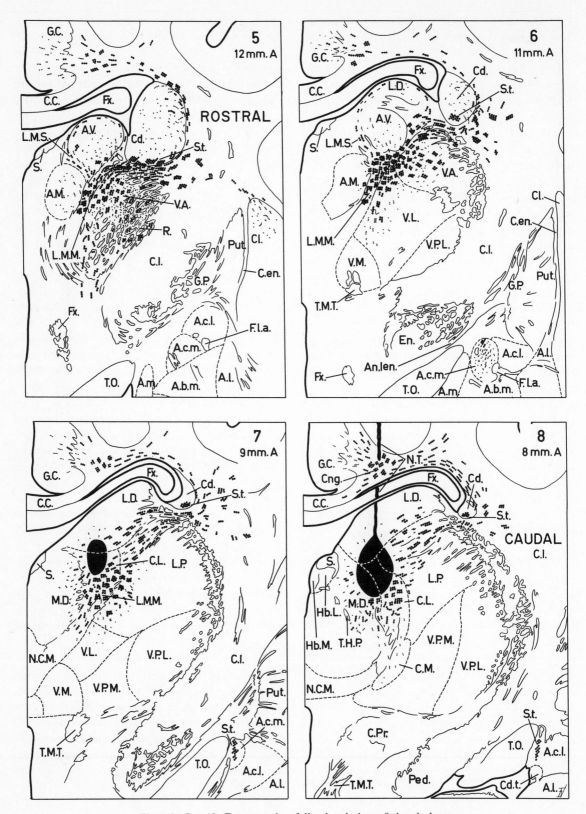

Fig. 16. Cat 15. Degeneration following lesion of the thalamus.

(L.M.M.) and the nucleus caudatus (Cd.), contains a greater number of coarse degenerating fibers.

In section 5 (Fig. 16) the same pattern of bifurcation of the degenerating fibers coursing in the lamina medullaris medialis (L.M.M.) can be observed; however, at this level the nucleus anteroventralis (A.V.) has gained the ventricular surface and, consequently, fibers of the lamina medullaris superficialis (L.M.S.) envelop the ventricular surface of this nucleus. A moderate number of terminal degenerating fibers enter the latter nucleus (sections 5 and 6, A.V.) and others, rather less abundant, penetrate the nucleus anteromedialis (sections 5 and 6, A.M.).

The fibers traversing the outer branch of bifurcation of the lamina medullaris medialis in sections 5 and 6 appear to be the same as those previously described in sections 7 and 8 coursing directly from the area of coagulation toward the capsula interna. It seems they form a continuous sheet of fibers whose rostral border is the outer branch of bifurcation of the lamina medullaris medialis.

One of the principal sites of termination of the fibers coursing through the thalamic laminae is the nucleus ventralis anterior (sections 5 and 6, V.A.); likewise, the nucleus reticularis (Fig. 15, section 4, and Fig. 16, section 5, R.) is pervaded by numerous, disintegrating, fine elements. Finally, the nucleus anterodorsalis (Fig. 15, section 4, A.D.) receives terminal degenerating fibers from the rostralmost laminar intrathalamic system.

Rostral to the thalamus (Fig. 15), the degenerating fibers shift laterally, gaining the medialmost part of the capsula interna, where they collect in large, transversally cut, bundles. These bundles, located between the head of the caudate (Cd.) and the capsula interna (C.I.), extend to rostral telencephalic levels (Fig. 15, sections 4 to 2). Their final destination will be considered after a description of the following group of fibers, interrupted by the lesion, which appear to course toward the amygdala following throughout the tail of the nucleus caudatus.

At rostral thalamic levels (Fig. 16, section 5) a small number of degenerating fibers enter the nucleus caudatus. These fibers follow caudally the same course as the stria terminalis (see in Fig. 16 S.t. within the tail of the caudatus in sections 5 to 8, and see also S.t. lateral to the tractus opticus, T.O., in the amygdaloid region, sections 6 to 8); their site of termination was located in the nucleus amygdaloideus centralis, pars medialis (section 6, A.c.m., and Fig. 18, *B*). No signs of degeneration were observed in other parts of the amygdala. The fascicle just described appears to be a rather small system of fibers closely associated with the tail of the caudate, entering the amygdala by its caudal aspect, and passing a little rostral to the levels where the tail of the caudate fuses imperceptibly with parts of the amygdaloid region.

It should be observed (Fig. 16, section 8) that the needle tract (N.T.) had cut through the cingulum (Cng.) and some fibers, coursing laterally to that bundle, enter the nucleus caudatus. These fibers form the ventral subcallosal radiations of the cingulum, which traverse the stratum subcallosum and enter the lamina zonalis which covers the limbic nuclei of the thalamus (Yakovlev and Locke, 1961); however, the latter fibers traverse the nucleus caudatus *en route* to the anterior thalamic nuclei and show remarkable differences in their trajectories with respect to the amygdalopetal fascicle just described. Although some fibers of the cingulum were seen entering the nucleus caudatus, they could not be traced in continuity with the stria terminalis, whereas at rostral thalamic levels (section 5), clearly, threads of degenerating fibers of the outer branch of bifurcation of the lamina medullaris medialis enter the nucleus caudatus (see also Fig. 17). No identifiable gross or histologic lesions other than those caused by the needle tract and by the area of coagulation were observed in any part of the brain.

In order to confirm the preceding observations, a group of animals were prepared by coagulation of the same region as in Cat 15. Unfortunately, these supplementary experiments failed to confirm this connection. In Cat 16 (not represented) the lesion involved a similar thalamic area to that in Cat 15, but severing of the cingulum also was present. The same pattern of degeneration

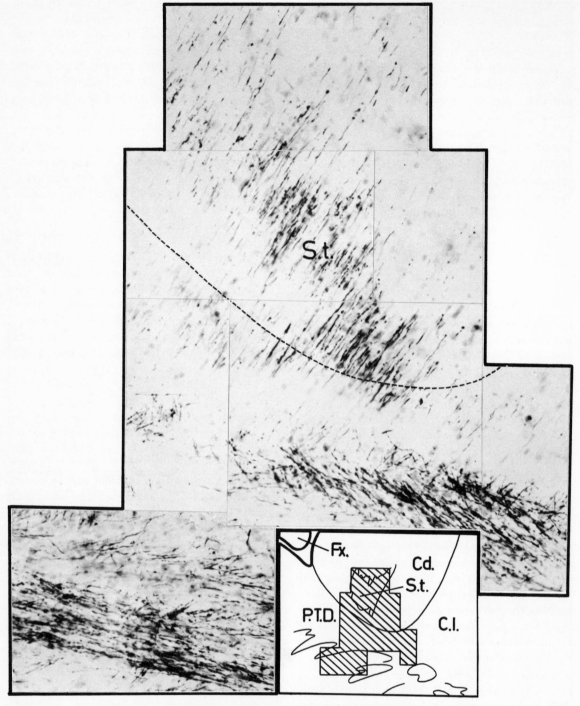

Fig. 17. Degenerating fibers entering the stria terminalis (S.t.) at the level of section 5 in Fig. 16. The diagram shows the area reconstructed photomicrographically. Nauta method.

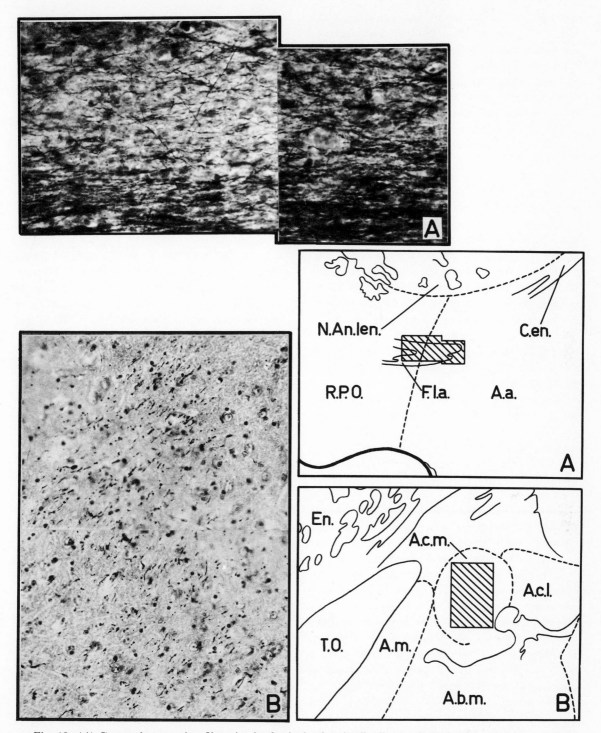

Fig. 18. (*A*) Coarse degenerating fibers in the fasciculus longitudinalis associationis after the lesion of Cat 1, taken from section 2 of Fig. 4. The diagram shows the location of the photomicrography. Nauta method.

(*B*) Fine preterminal and terminal degenerating fibers in the nucleus amygdaloideus centralis, pars medialis, after the lesion of Cat 15. The diagram shows the area reproduced in the photomicrography. Nauta method.

was observed: fibers of the outer branch of bifurcation of the lamina medullaris medialis enter the nucleus caudatus and could be followed backward, via the stria terminalis, to the nucleus amygdaloideus centralis, pars medialis but fibers of the ventral cingulum entering the nucleus caudatus also were observed.

In Cat 20 (not represented) a contralateral approach was intended in order to spare the cingulum, but the needle tract caused damage to the corpus callosum and fornix and no additional observations could be added. Cats 22 and 23 were not useful for any additional observations.

The problem whether or not the fibers traced from the intralaminar cell groups of the thalamus to the amygdala via the stria terminalis represents a direct connection cannot be decided on the basis of the present experimental material. Later, in sagittal Golgi sections of the rat, fibers will be demonstrated coursing rostrally through the thalamus to bend dorsally at the level of the bed nucleus of the stria terminalis and to enter the stria, where they follow backward toward the amygdala (Fig. 56).

The rostral continuations of the degenerating fibers traced in Cat 15 through the thalamic laminae enter the medial stratum of the capsula interna. Located along the lateral border of the caudatus (Fig. 15, sections 4 to 2), they course rostrally (section 1) and bifurcate. One group approaches the capsular edge below the gyrus sigmoideus (G.Sg.), where degenerating fibers are present in several fascicles made up of coarse elements; these terminate as numerous fine fibers in the layers of the gyrus sigmoideus.

The other system of fibers, rather less abundant, approach the gyrus proreus (G.Pr.) and adjoining lateral parts of the capsula interna (C.I.); a moderate number of degenerating fibers of terminal character distribute themselves through the deep layers of the cortex of the gyrus proreus (G.Pr.), and some scanty terminal degeneration appears to enter the gyrus orbitalis (G.Orb.).

The group of coarse fibers traveling between the nucleus caudatus and the capsula interna (sections 2 to 4) furnishes abundant terminal degeneration to the head of the caudate. The fibers appear to enter the caudate all along the entire border that separates it from the capsula interna and, although diffusely distributed throughout its extent, some terminal fibers concentrate in the ventral region of the head of the caudate in the bed nucleus of the stria terminalis (Fig. 15, section 3, B.S.t.). A few coarse fibers turn ventrally, pierce the commissura anterior, pars temporalis (C.A.p.t.), and furnish scanty terminal degeneration to the nucleus ansae lenticularis (N.An.len.) and farther rostrally to the deep layers of the tuberculum olfactorium (section 2, Tu.Olf.). The scarce degeneration traced through the ventromedial parts of the capsula interna to the globus pallidus (section 4, G.P.) appear to be from the same group.

From the lateral border of the thalamo-telencephalic degenerating system described in the present experiment, another stream of fibers turns ventrolaterally, pierces the capsula interna, and enters the claustrum (sections 2 to 5, Cl.) along its dorsal border, where a moderate number of terminal degenerating fibers are distributed through its dorsal region. Other fibers follow the capsula externa (section 3, C.en.) and finally also enter the claustrum (section 2).

From the lateral border of the head of the nucleus caudatus, a group of fibers turns dorsomedially (Fig. 15, sections 4 to 2), entering the gyrus cinguli (G.C.), where the fibers become intermingled with the degenerating ones observed in this gyrus at more caudal levels (sections 8 to 4). Undoubtedly, this latter group of fibers, which dorsally surround the head of the nucleus caudatus and enter the gyrus cinguli (G.C.), are intermixed with degenerating fibers caused by the needle tract. The fibers observed in the gyrus sigmoideus (section 1, G.Sg.) probably also represent degeneration of the cingulum.

CAT 17

The lesion of this animal was placed in the capsula externa with encroachment upon the nucleus amygdaloideus lateralis where the dorsal edge of the latter lies between the putamen and claustrum. Some involvement of the two latter nuclei also was observed (Fig. 3, sections 11 to 15).

The brain was studied in horizontal sections.

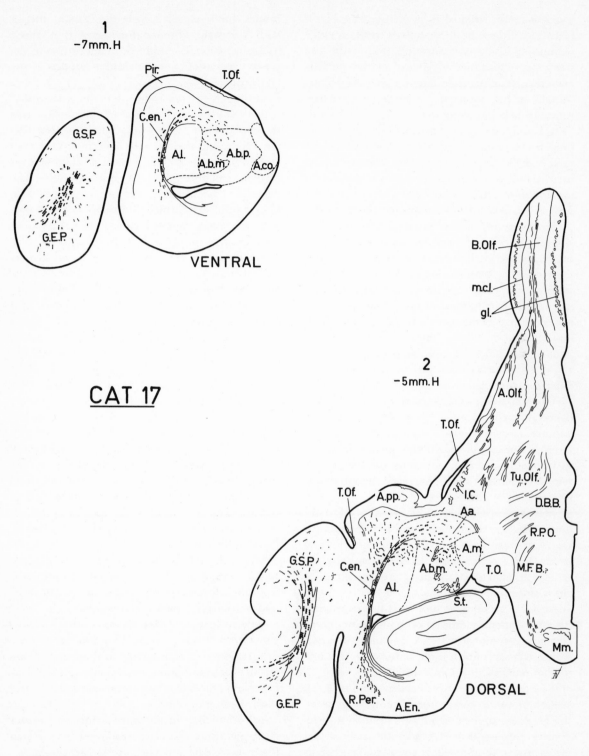

Fig. 19. Cat 17. Degeneration following lesion of the capsula externa and adjoining putamen, claustrum, and amygdaloid complex, as indicated in Figs. 20 and 21. Horizontal sections.

The lesion was situated in a narrow zone like a pedicle, which joins the gyri sylvianus, ectosylvianus, and suprasylvianus posterior with the more rostral and medial parts of the brain. This pedicle is flanked rostrolaterally by the fissura rhinalis and caudomedially by the wall of the temporal horn of the third ventricle (Figs. 20 and 21, sections 3 to 5). At this level (1 to 3 below the horizontal zero plane), the axis of the pedicle is formed by the capsula externa (C.en.) as a rostral prolongation of the white stalks of the three gyri just mentioned. The lesion interrupted numerous fibers coursing through this pedicle to and from the temporal gyri.

Four systems of degenerating fibers can be traced in this experiment: (a) cortical fibers through both capsulae externa and extrema; (b) a commissural component to the contralateral hemisphere; (c) fibers to the amygdaloid complex; and (d) fibers to several regions of the di- and mesencephalon by way of the ansa lenticularis and pedunculus thalami extracapsularis.

(a) The lesion interrupted many fibers that enter the white matter of the gyri sylvianus posterior (G.S.P.), ectosylvianus posterior (G.E.P.), and suprasylvianus posterior (G.Su.P.) through the capsula externa (Figs. 19, 20 and 21, sections 1 to 6). Abundant fine degenerating fibers leave the white temporal matter and enter the gray of the temporal convolutions. Some reach the superficial stratum of the cortex.

A few coarse disintegrating fibers travel in the periventricular white matter and terminate in the regio perirhinalis (sections 2 to 6, R.Per.). Rostral to the lesion coarse fiber degeneration could be traced through the capsula externa as far as to the gyrus proreus (sections 4 to 6, G.Pr.). Along this course, a moderate number of fine degenerating fibers enter the deep stratum of the area praepiriformis (sections 2 to 4, A.pp.) and, more dorsally (sections 5 and 6), the putamen (Put.) and medial parts of the claustrum (Cl.).

Just dorsal to the area of coagulation (Fig. 21, section 6), condensed groups of coarse degenerating fibers are present. From here abundant disintegrating elements skirt the caudal pole of the claustrum (Cl.) and follow the capsula extrema (C.em.) distributing a moderate number of terminal fibers to the gyrus sylvianus anterior (G.S.A.), gyrus ectosylvianus anterior (G.E.A.), and gyrus orbitalis (G.Orb.); medially scanty terminal elements enter the lateral aspect of the claustrum (Cl.).

Although the possibility that the lesion interrupted some ascending fibers from the di- and mesencephalon cannot be excluded, it is clear that we are dealing mainly with cortico-cortical association pathways coursing to several homolateral convolutions throughout the outer capsules. The posterior limb of the capsula interna (section 5, C.I.p.l.), through which the temporal lobe projects mainly to caudal levels of the brain, is almost entirely free from degenerating fibers.

(b) Rostral to the lesion (Figs. 20 and 21, sections 4 and 5), a massive group of degenerating fibers, initially interwoven with fibers of the capsula externa, enters the commissura anterior, pars temporalis (C.A.p.t.), crosses the midline, and is distributed in the contralateral temporal convolutions, although not so profusely as homolaterally.

(c) The ventralmost transit of degenerating fibers follows rostrally through the capsula externa (Fig. 19, section 2, C.en.) until it spreads out in the poorly defined area amygdaloidea anterior (section 2, A.a.), where a moderate number of fine degenerating fibers are present. From here scanty terminal fibers turn backward and enter the rostral third of the nucleus amygdaloideus basalis, pars magnocellularis (sections 1 and 2, A.b.m.) and of the nucleus amygdaloideus lateralis (section 2, A.l.).

The amygdala is also supplied diffusely by terminal degenerating fibers, which extend medially from the capsula externa to the nucleus amygdaloideus centralis, pars lateralis (section 3, A.c.l.) and the nucleus amygdaloideus lateralis (A.l.), and rostrally to the area amygdaloidea anterior (A.a.), in a region just rostrolateral to the compact fasciculus longitudinalis associationis (F.l.a.). This latter, on the other hand, remains entirely free from degenerating fibers.

(d) The lesion in this animal furnishes a very clear picture of the course followed by the ansa lenticularis (An.len.). The latter, in horizontal sections, can be considered to be formed by two seg-

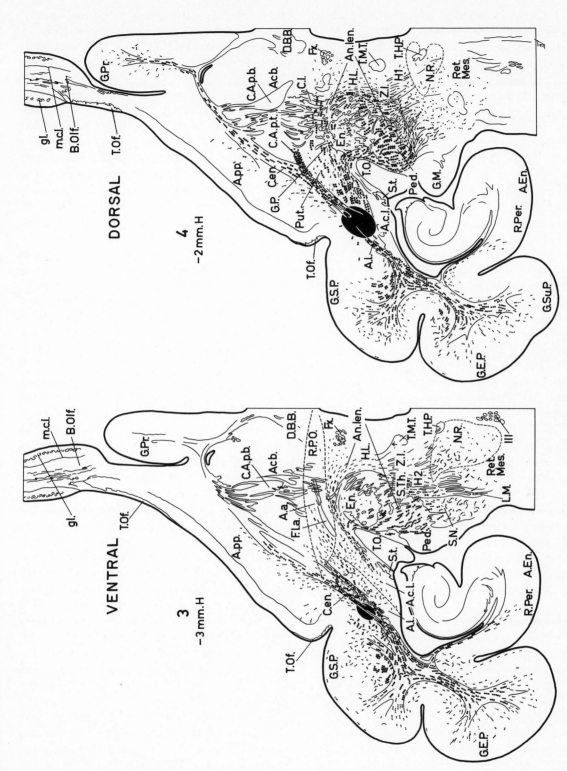

Fig. 20. Cat 17. Degeneration following lesion of the capsula externa and adjoining putamen, claustrum, and amygdaloid complex, as indicated in this figure and Fig. 21. Horizontal sections.

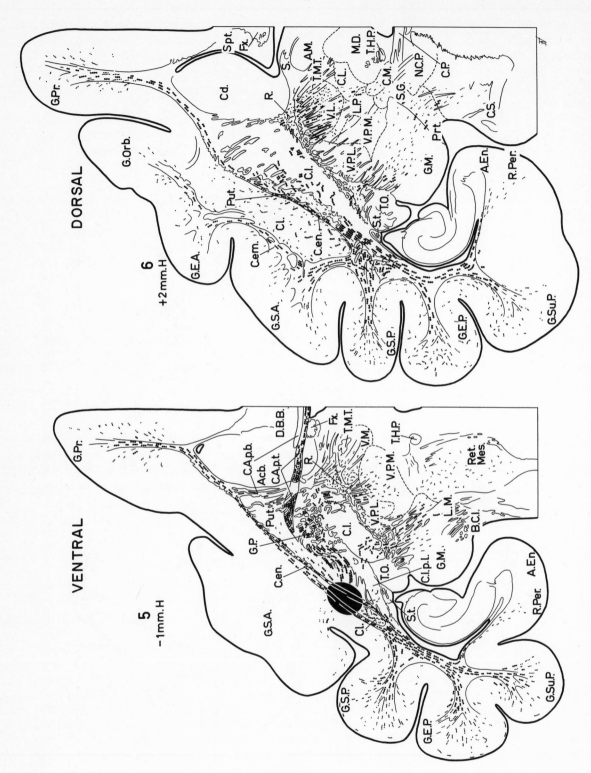

Fig. 21. Cat 17. Degeneration following lesion of the capsula externa and adjoining putamen, claustrum, and amygdaloid complex, as indicated in this figure and Fig. 20. Horizontal sections.

ments that meet, forming an angle embracing the tractus opticus (Fig. 20, section 4, T.O.); the first segment is made up of compact fascicles of coarse degenerating fibers which extend from the lesion to the nucleus entopeduncularis (section 4, En.). Part of this segment courses through the caudal edge of the putamen (section 4, Put.) and alongside the medial border of the nucleus amygdaloideus centralis, pars lateralis (section 3, A.c.l.). From the level of the lateral border of the nucleus entopeduncularis large numbers of degenerating fibers curve caudally, forming the second segment of the ansa lenticularis, which spreads out in fanlike fashion so that at least three divisions can be followed. The rostral border of this fanlike portion swings around the nucleus entopeduncularis, leaving a moderate number of fine degenerating terminal fibers just lateral to the fornix (section 3, Fx.). A very few terminal fibrils appear to enter the regio praeoptica (R.P.O.); other fibers turn backward, coursing through the hypothalamus lateralis (sections 3 and 4, H.L.). A second, rather scanty, part of the degenerating fibers pierces diffusely the nucleus entopeduncularis (Fig. 20, sections 3 and 4, En.), where a few fibers terminate, while others interweave with those of the pedunculus cerebri (Ped.). The third division forms the lateral edge of this fanlike part of the ansa lenticularis and courses just along the medial edge of the tractus opticus (Fig. 20, sections 3 and 4, T.O.).

Just behind the pedunculus cerebri this fantail system of the ansa lenticularis collects along the medial border of the former, furnishing numerous terminal fibers to the zona incerta (Z.I.), the substantia nigra (S.N.), the H1 and H2 fields, and the nucleus subthalamicus (S.Th.); still more caudally abundant fine terminal degeneration can be traced to the mesencephalic reticular formation (Ret.Mes.) and the nucleus ruber (N.R.).

At more dorsal levels, other fibers (Fig. 21, sections 5 and 6) follow wavy trajectories and project to the globus pallidus (section 5, G.P.). Coarse degenerating fibers interweave with fibers of the capsula interna (C.I.), and terminate in the nucleus reticularis thalami (R.). Fibers of this system pierce the capsula interna at different points (sections 5 and 6, C.I.) and enter various thalamic nuclei. From the genu of the capsula interna (Fig. 21) caudally a stream of fine degenerating fibers enters the thalamic nuclei: ventralis medialis, ventralis posteromedialis (section 5, V.M. and V.P.M., respectively), and ventralis lateralis (section 6, V.L.). Some extend on through the centralis lateralis (C.L.) and lateralis posterior (L.P.; a few terminate here), to end diffusely among cells of the nuclei centrum medianum (C.M.) and suprageniculatus (S.G.) and adjoining regions of the nuclei lateralis posterior (L.P.) and ventralis posteromedialis (V.P.M.).

Other fibers swing caudomedially around the tractus opticus (Fig. 21, sections 5 and 6, T.O.); some fibers terminate in the nucleus geniculatus medialis (G.M.), while others extend to the mesencephalic reticular formation (section 5, Ret. Mes.), pretectal region, and adjoining part of the colliculus superior (section 6, Prt. and C.S., respectively), where they terminate.

The course outlined by the fibers entering the thalamic nuclei, just discussed, coincides with the trajectory of the pedunculus thalami extracapsularis of Klingler and Gloor (1960).

CAT 19

Three cats (Nos. 18, 19, and 25) were prepared by coagulations in the gyrus orbitalis for the study of the projection pathways from this cortical area to the amygdala. The degeneration traced in these three animals was strikingly similar. Cat 19 will be described as representative of these experiments.

As shown by Fig. 3, sections 1 and 2, and Fig. 22, sections 2 and 3, the lesion occupies a large area of the lateral aspects of the gyrus orbitalis without involvement of the white matter. In section 3, where the area of coagulation attains its greatest extension, the defect involves the entire thickness of the gray matter between the white stalk of the gyrus orbitalis and the sulcus orbitalis, which separates caudally the former from the gyrus ectosylvianus anterior (G.E.A.).

Massive fiber degeneration extends from the lesion to the white stalk of the gyrus orbitalis (Fig. 22, sections 1 to 3). From here rostrally three threads of degenerating fibers enter the

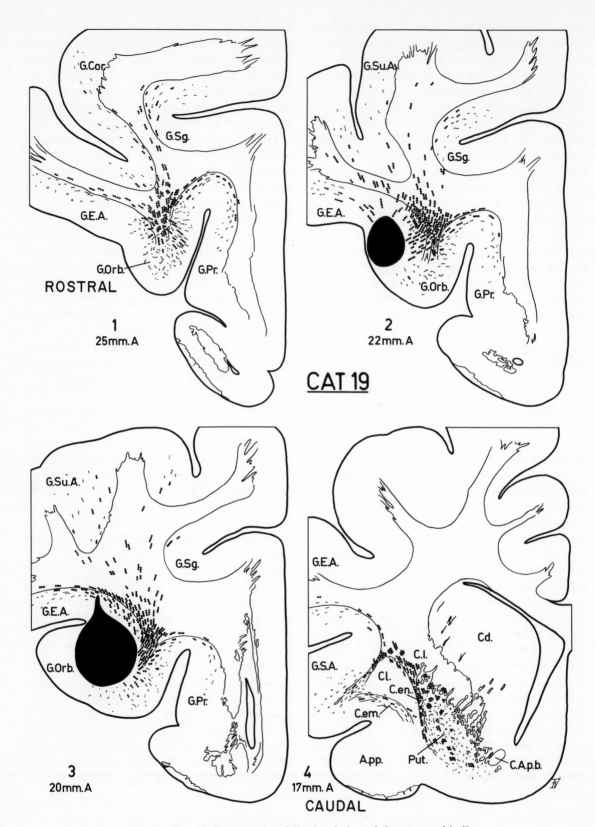

Fig. 22. Cat 19. Degeneration following lesion of the gyrus orbitalis.

white matter of the brain; one of them takes a lateral direction through the white matter of the gyrus ectosylvianus anterior (sections 1 to 4, G.E.A.), scarce terminal fibrils enter the gray of this gyrus. Other coarse degenerating fibers ascend dorsally and supply a moderate number of fine degenerating fibers to the gyrus coronalis (section 1, G.Cor.), gyrus suprasylvianus anterior (sections 2 and 3, G.Su.A.), and gyrus sigmoideus in the depth of the sulcus cruciatus (sections 1 and 2, G.Sg.), while a third stream, following the limits between the gray and the white matter, reaches the dorsal half of the gyrus proreus (sections 1 to 3, G.Pr.).

From the area of coagulation caudally the bulk of coarse degenerating fibers ride upon the dorsal limits of the claustrum (Fig. 22, section 4, Cl.), entering the outer capsules and the white matter separating the lateral face of the claustrum from the gray of the gyrus sylvianus anterior (G.S.A.). The putamen (section 4, Put.) appears pervaded by numerous disintegrating fibers of various calibers. Fine fibers disperse diffusely in this nucleus, while coarse fibers appear, grouped in several bundles. From this level three major groups of degenerating fibers can be followed caudally: (*a*) one of them courses throughout the capsula externa; (*b*) a second system follows the rostrocaudal extension of the limits separating the nucleus lenticularis from the amygdaloid complex; (*c*) a third group occupies part of the capsula interna, where it can be followed to the nucleus ventralis posterolateralis of the thalamus.

(*a*) The system of fibers coursing in the capsula externa extends alongside the lateral face of the nucleus amygdaloideus lateralis (Fig. 23, sections 5 to 8, A.l.). Numerous fine degenerating fibers enter that nucleus, while more ventrally coarse, as well as fine, fibers enter the ventral region of this amygdaloid nucleus and the pars parvocellularis of the nucleus amygdaloideus basalis (sections 6 and 7, A.b.p.) where the capsula externa (C.en.) swings around the ventral border of the amygdala, separating the latter from the underlying cortex piriformis (Pir.).

(*b*) A second system of fibers, as mentioned before, extends through the limits separating dorsally the amygdala from the nucleus lenticularis.

This system of fibers, composed of several fascicles made up of coarse degenerating fibers, was cut transversally in sections 5 and 6 (Fig. 23). From that level numerous degenerating fine fibers descend toward the area amygdaloidea anterior (section 5, A.a.) and, more caudally, to the nuclei centralis (A.c.l. and A.c.m.), basalis (A.b.m. and A.b.p.), and medialis (A.m.) of the amygdala (sections 6 to 8). At chiasmatic levels (section 6), the entire area of the amygdala is filled with numerous degenerating fine fibers which reached the complex either via capsula externa or via the sublenticular degenerating system.

The most medial sublenticular part of this second system of fibers enters the nucleus amygdaloideus centralis, pars medialis (Fig. 23, section 7, A.c.m.), and collects just lateral to the tractus opticus (T.O.). These fibers enter the stria terminalis (Fig. 23, section 8, S.t., lateral to the tractus opticus, T.O.) and, following the same course as the rest of the amygdalofugal stria fibers (sections 7 and 8, S.t., within the nucleus caudatus Cd.), reach the bed nucleus of the stria terminalis (sections 5 and 6, B.S.t.), where a moderate number of terminal fibers end. Part of this stria component enters the inferior stratum of the commissura anterior (section 5, C.A.) and crosses to the contralateral stria bed, where it could not be followed farther.

This cortical stria component was also observed in Cats 18 and 25, with similar lesions in the gyrus orbitalis (not represented), but it is difficult to admit the existence of a neocortical component of the stria terminalis; the possibility that a vascular lesion affected the stria terminalis needs to be taken into account.

(*c*) The third system of fibers represents a connection established between the gyrus orbitalis and the nucleus ventralis posterolateralis of the thalamus. From the group of degenerating bundles riding upon the dorsomedial aspect of the claustrum (Fig. 22, section 4, Cl.) a moderate number of coarse degenerating fibers swing medially, enter the capsula interna (Fig. 23, sections 5 to 8, C.I.), and, broken up into small fascicles consisting of coarse fibers, reach almost the entire extension of the nucleus ventralis posterolateralis (section 8, V.P.L.), where they terminate.

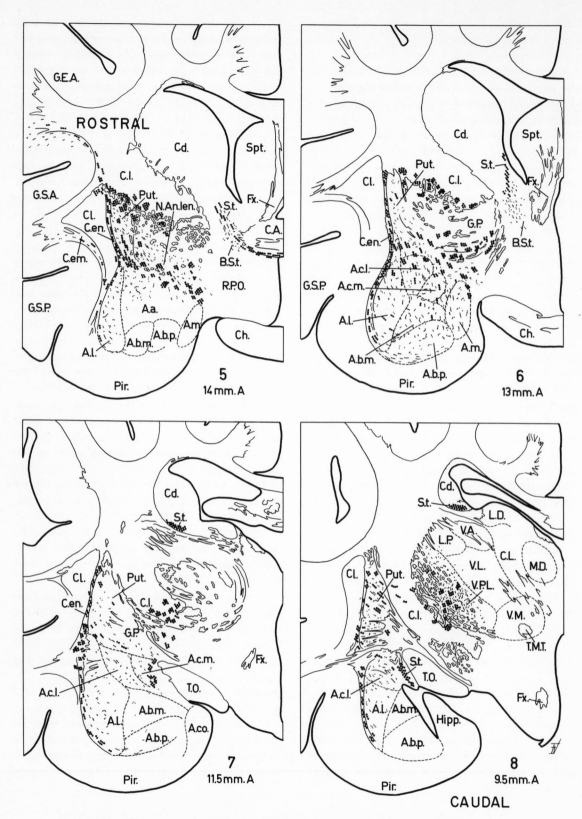

Fig. 23. Cat 19. Degeneration following lesion of the gyrus orbitalis, as indicated in Fig. 22.

As shown in Fig. 3, sections 10 to 12, and Fig. 24, section 4, in this case the lesion was situated in the dorsal part of the amygdaloid complex, involving parts of the nucleus amygdaloideus centralis (A.c.m. and A.c.l.) and medial aspects of the nucleus amygdaloideus lateralis (A.l.). The area of coagulation spared the compact part of the stria terminalis (section 4, S.t.), whereas the fasciculus longitudinalis associationis (F.l.a.) was involved in part by the lesion.

Five systems of projections were traced from the lesion: (*a*) diffuse projections, in part through the outer capsules, to the putamen, claustrum, and gyrus sylvianus posterior; (*b*) fibers coursing rostrally to the regio praeoptica and area amygdaloidea anterior through both compact and diffuse components of the rostral projection system of the amgydala; (*c*) a commissural component of the stria terminalis; (*d*) fibers coursing medially through the ansa lenticularis; and finally (*e*) diffuse fine degenerating fibers to part of the remainder of the amygdaloid nuclei.

(*a*) From the lateral aspects of the lesion abundant coarse degenerating fibers traverse in various directions the nucleus amygdaloideus lateralis (Fig. 24, sections 3 and 4, A.l.). Part of these fibers descend ventrally and enter the capsula externa (C.en.), taking a medial direction in the zone separating the amygdala from the underlying cortex piriformis (Pir.). It appears that these fibers end in the deep layers of the latter.

Another group of coarse fibers collect in the capsula externa in the base of the claustrum just in front of the hilus of the gyrus sylvianus posterior (sections 3 and 4, G.S.P.). From here a moderate number of coarse, as well as fine, degenerating fibers enter the white matter of this gyrus and distribute themselves through its deep gray layers.

A third group of fibers ascend diffusely through the lateral aspects of the putamen (sections 3 and 4, Put.), claustrum (Cl.), and deep layers of the gray lateral to the latter. Part of these fibers follow rostral trajectories through the capsula externa, reaching more rostral parts of the putamen and claustrum (sections 1 and 2, Put. and Cl., respectively).

(*b*) The degeneration traceable through both compact and diffuse parts of the rostral projection system of the amygdala appears similar to that traced in Cats 1, 3, 5, and 6. The compact part of this system, that is, the fasciculus longitudinalis associationis, shows massive fiber degeneration through the amygdala (Fig. 24, sections 3 and 4, F.l.a.). At preoptic levels (section 2, F.l.a.) it takes a medial direction, spreading out in the lateral parts of the regio praeoptica (R.P.O.), where a moderate number of fine degenerating fibers disperse among the cells of this region. A few fibers appear to enter the pedunculus thalami inferior (P.T.I.), but they could not be followed to any nuclear group.

The termination of the diffuse or lateral part of the rostral projection system of the amygdala is clearly shown at rostral amygdaloid levels in the area amygdaloidea anterior (section 2, A.a.). Scattered fine degenerating elements are distributed among the cell groups of this region, and still more rostrally (section 1) a few fine degenerating fibers are present in the lateral area of the regio praeoptica (R.P.O.) and the medial part of the area amygdaloidea anterior (A.a.).

(*c*) In this experiment the lesion caused damage in the commissural component of the stria terminalis. Within the amygdala, part of the stria appears degenerated (Fig. 24, section 4, S.t.). At commissural levels (section 1) the degenerating fibers of the stria descend directly through its bed nuclei (B.S.t.) to the commissura anterior (C.A.). Part of these fibers, presenting terminal characteristics, appear to end among cells of the bed nucleus, whereas a minimal number enter the commissura and follow contralaterally through its dorsal stratum. The contralateral bed nuclei were entirely free from degenerating elements. No degeneration was traceable to any contralateral structure. This might furnish evidence that these commissural fibers synapse with dendrites of the gray matter protruding into the commissura anterior, as was suggested with Cats 3 and 6.

(*d*) A few fine degenerating fibers skirt the dorsolateral edge of the tractus opticus (Fig. 24, sections 3 and 4, T.O.) and enters the ansa lenticularis (An.len.), but they could not be followed farther.

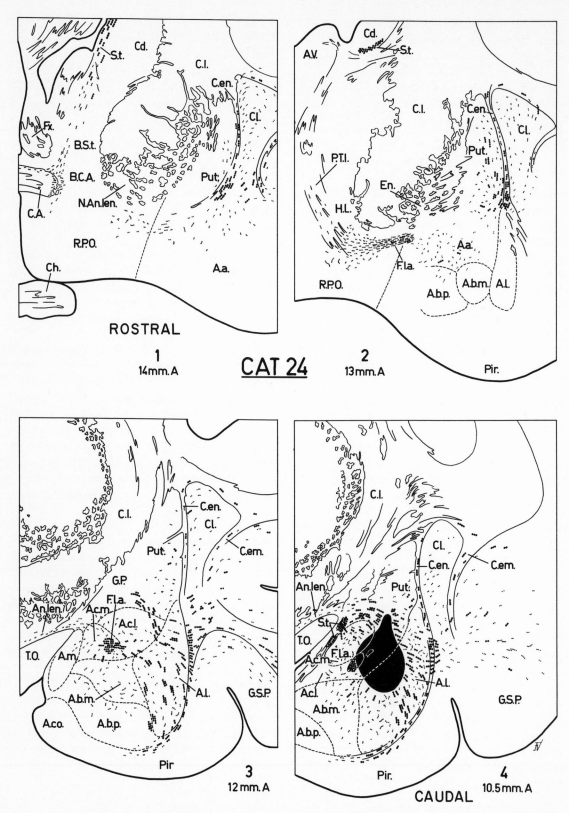

Fig. 24. Cat 24. Degeneration following lesion of the amygdaloid complex.

(*e*) The amygdaloid nuclei appear to be pervaded diffusely by numerous coarse, as well as fine, degenerating fibers. These fibers, originated from damage to collateral axonal branches or to intermediate axoned cells of the amygdala, extend to the nuclei lateralis (Fig. 24, sections 3 and 4, A.l.), both subdivisions of the centralis (A.c.l. and A.c.m.), and the basalis (A.b.m. and A.b.p.). Within the last the degeneration is more abundant in its magnocellular division (A.b.m.). Only a few fine degenerating fibers enter the nucleus amygdaloideus medialis (section 3, A.m.), while the nucleus amygdaloideus corticalis (A.co.) is entirely free from degeneration.

3   Golgi Observations

Since Golgi (1875) applied his own method to the study of the mammalian olfactory bulb, earlier workers (Cajal, 1890; van Gehuchten and Martin, 1891; Kölliker, 1896) and recently others (Clark and Warwick, 1946; Allison and Warwick, 1949; Allison, 1953; Clark, 1957) have examined this structure in detail.

With the triple-impregnation procedure of the Golgi method described in the present work, I have been able to observe certain new features not seen in routinely stained Golgi sections. Figure 25 is a horizontal section through the olfactory bulb of a 9-day-old albino rat. In the layer of glomeruli (layer II) I drew the olfactory fibrillar component of three glomeruli (2, 4, and 12). Fibers entering from layer I (layer of olfactory fibers) to form the glomerular plexus are unmyelinated and, as clearly seen in glomerulus 2, they proceed from a short number of main branches (see, for example, fiber 17) which divide, usually at some distance from the glomerulus, into many secondary branches like a brush. It should be noted that each glomerulus is formed by several brushlike endings, and not of individual elements.

Another component of the glomerulus is represented by the dendritic branches of the large mitral cells. It has been assumed for a long time that each mitral cell possesses a single dendritic process ending in a brush of short branches spreading within a glomerular plexus. Repeated observations showed me the existence of a moderate number of mitral cells with two or three descending dendrites for one or more glomeruli; thus, there may be mitral cells, such as e in Fig. 25 and that of Fig. 26B, possessing two dendritic branches (2 and 3) whose extremities arborize within a single glomerulus, and others, such as d in Fig. 25, having three dendrites for three different glomeruli (5, 6, and 7). Another example is furnished by cell f, possessing two dendrites which, by means of their terminal brushes, form the glomeruli 11 and 15.

A third component of the olfactory glomeruli is represented by the short-axoned periglomerular cells. Cell k (Fig. 25) possesses three main dendritic branches forming part of the three glomeruli 8, 9, and 12. A special axo-dendritic contact was observed when fiber 22 approached one of the dendrites of cell k. This fiber divides into several sprouts which climb along the dendrite. The axon of cell k (labeled 1k) follows a tortuous course within the layer of glomeruli, giving off several collaterals among other glomeruli. Another example of a periglomerular cell is that labeled l, which possesses a dendritic tree intermingled with the final branches furnished by the large dendrite of the tufted cell j. The axon of cell l behaves in the same manner as that of cell k. The periglomerular or external granule cells are said to connect different glomeruli within this layer. It is interesting to point out that Cragg (1962) has traced in the rabbit fine degenerating fibers from the cortex praepiriformis to the glomerular level in the olfactory bulb, ending in relation to the external granule cells. Fiber 22 in Fig. 25 might represent such centrifugal olfactory-bulb fibers.

Beneath the glomeruli, the layer III or layer of tufted cells appears as a poorly defined region with scarce fibers and tufted cells disposed randomly. This layer contains mainly the vertical, sometimes oblique, main dendritic branches of

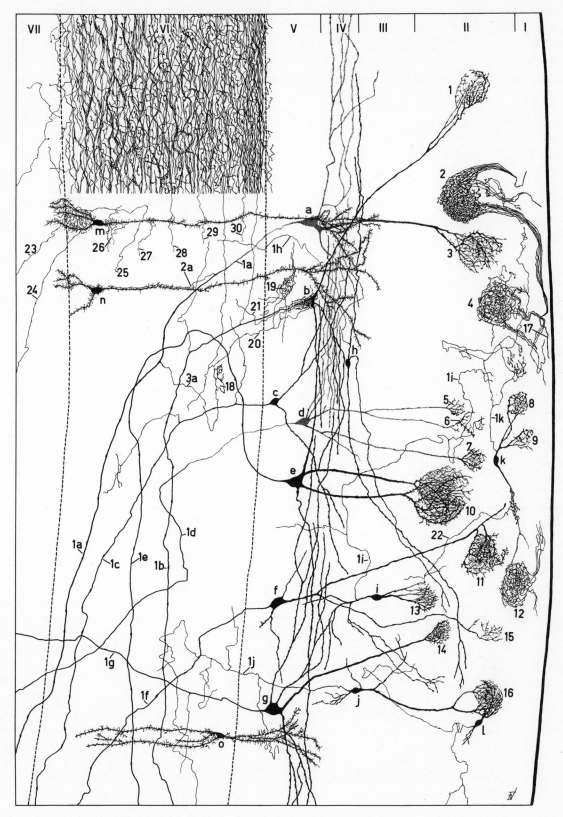

Fig. 25. Rat 43R, albino, 9 days old. Horizontal section through the olfactory bulb. Details of the fibrillar and cellular structure. Golgi method.

mitral cells. The tufted cells have fusiform bodies (cells *i* and *j*) and, like mitral cells, one or more dendrites coursing toward the glomerular layer. Although not so profuse as the final brushes of dendrites of mitral cells, these dendrites participate also in the construction of glomeruli. The axons of tufted cells are of smaller caliber than those of mitral cells (compare the mitral axon *1f* with the tufted one *1j*) and it should be pointed out that after repeated observations on numerous Golgi sections of the olfactory bulb sectioned in different planes and, likewise, after careful examination of Cajal's original preparations[1] I never could follow the axons of tufted cells beyond the limits of the olfactory bulb. The axon *1i* of the tufted cell *i* in Fig. 25 follows a wavy horizontal course below the layer of mitral cells until it penetrates among the glomeruli. Other observations showed some final branches of tufted-cell axons entering the glomeruli. Conversely, the axon *1j* of tufted cell *j* ascends to the internal plexiform layer (layer VI), where it divides, giving off several collaterals contributing to the formation of the dense plexus of this layer.

From these observations tufted cells may be referred to as *inwardly displaced periglomerular cells,* and not as *outwardly displaced mitral cells* as is generally believed. Now the question arises what kinds of cells give origin to the projection pathways of the olfactory bulb, that is, the tractus olfactorius lateralis and the commissura anterior, pars bulbaris? This question will be discussed below.

The external plexiform layer, or layer IV, is composed almost exclusively of the long horizontal dendrites of mitral cells intimately intermingled with the descending expansions of the internal granular cells. Rarely, one finds small horizontal cells, such as *h*, with very long dendrites and possessing a thin axon (*1h*) ascending toward the internal plexiform layer. In the case of cell *h* two long dendrites were observed descending obliquely to the layer of glomeruli but without establishing relations with them.

The layer of mitral cells, or layer V, appears as a row of large, triangular bodies with apexes pointing inward. The mitral bodies, in well-impregnated Golgi sections, frequently appear enveloped by a dense fibrillar network furnished by fine fibers of the internal plexiform layer (layer VI); see, for example, the nest formed by fibers *20* and *21* around mitral cell *b* and nest *19* in which the cell body was not stained (Fig. 25).

In disagreement with Allison (1953), who states that the axons of mitral cells are thin and difficult to follow, these axons are the thickest ones I have observed in the brain. Several times I could follow in a Golgi horizontal section 200 μ thick a single mitral axon as far as the fibrillar layer of the periamygdaloid cortex. The caliber of the mitral axons is illustrated in Fig. 26*A* (see the axon of mitral cell *b*) and in *B* of the same figure (axon 1). The axons of mitral cells enter directly the internal plexiform layer, emitting several collaterals here to form part of the dense fibrillar network of this layer. Several examples of these collaterals are given in Fig. 25. Axon *1a* of mitral cell *a* emits a collateral (*2a*) which approaches the descending expansion of the internal granular cell *n* and climbs along it, giving off many short sprouts to this process. The collateral *3a* of the same mitral-cell axon forms a nest at *18*. According to Cajal (1890) and Allison (1953), there are recurrent collaterals of the axons of mitral cells to the mitral cell bodies and external plexiform layer. Herrick (1924b) pointed out the existence of interneurons between axon collaterals and mitral-cell bodies.

Finally, one finds mitral cells (cell *c*) devoid of descending glomerular dendrites. These cells are rare and they appear slightly displaced inward with respect to the rows of true mitral cells.

Under low magnification the internal plexiform layer (layer VI of Fig. 25) appears as a dense plexus formed by small bundles of fibers of various calibers following wavy trajectories and leaving ovoid spaces in which the internal granular cells are located. The plexus is formed by the axons and its collaterals of the mitral cells, the axons of tufted cells, and extrinsic fibers coming from outside the olfactory bulb.

Some of the most interesting components of the internal plexiform layer are the internal granular cells. These cells have bodies of medium size and ovoid shape lacking expansions of axonal appearance. While Kölliker (1896) believed these

[1]From the Cajal Museum at the Instituto Cajal, Madrid.

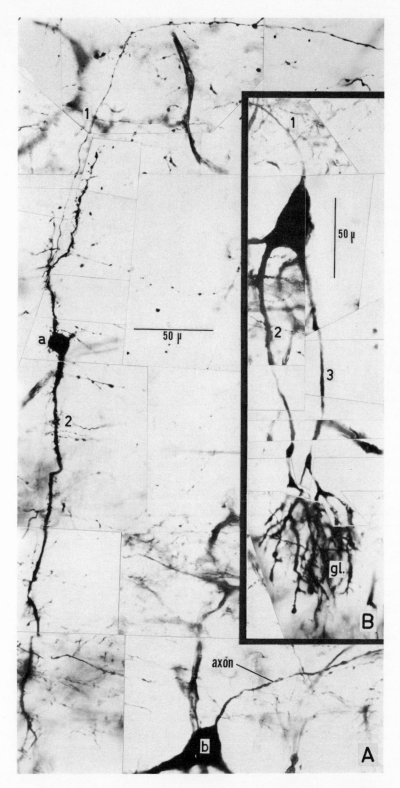

Fig. 26. Rat 43R, albino, 9 days old. (*A*) Mosaic photomicrographic reconstruction of an internal granular cell (*a*) of the olfactory bulb and some of its afferent fibers. Presumed axo-axonal contacts occur where fibers of the internal plexiform layer cross over the descendent expansion at point *2*. Cell *b* is a mitral cell. Golgi method.

(*B*) Mosaic photomicrographic reconstruction of a mitral cell of the olfactory bulb provided with two descending dendrites (*2* and *3*) forming a single glomerulus (*gl.*). Golgi method.

elements to represent a variety of neuroglia, Cajal (1890) considered them true neurons, similar to the amacrine cells of the retina. The descending expansion would play the role of an axon even though morphologically its appearance is far from this homology. The internal granular cells possess a brush of dendrites pointing toward the ventricular cleft of the olfactory bulb; the length of these dendrites depends on the location of the cells in the layer (compare cells *m* and *n* with cell *o* in Fig. 25). The descending expansion arises from the outer pole of the cell and reaches the layer of mitral cells, where it divides into several secondary branches spreading out in the external plexiform layer (layer IV). The inner dendritic brush, the body, and the descending expansion appear constantly covered by numerous fine sprouts as well as by delicate end ramifications. In well-impregnated Golgi material the numerous sprouts covering these cells are attached to thin fibers. This is the case of fibers *23–26* with respect to cell *m* in Fig. 25. Fibers of the internal plexiform layer cross over the descending expansion of the internal granular cells and emit delicate twigs making synaptic contacts on these expansions (Fig. 25, fibers *27–30;* Fig. 26*A,* group of fibers *2* over the descending expansion of cell *a*).

The descending expansions of the internal granular cells function as true axons in the sense that they transmit the activity of the internal plexiform layer to the horizontal dendrites of mitral cells. With this point of view the synapses established by fibers *26–30* and *2a* on granular cells *m* and *n* of Fig. 25 and the group of fibers *2* on the descending expansion of cell *a* in Fig. 26*A* might be considered as axo-axonic synapses. Other studies are necessary to fully verify these observations.

As will be described later, my Golgi studies reveal that in the rat fibers of the commissura anterior, pars bulbaris, originate largely in the pyramidal cells of the nucleus olfactorius anterior. The axons of these cells give off a thin collateral, which enters the internal plexiform layer of the homolateral olfactory bulb, and one main branch, which courses throughout the pars bulbaris of the commissura anterior to the contralateral anterior olfactory nucleus and olfactory bulb.

It is postulated that the direct influence exerted by the axons of cells of the anterior olfactory nucleus upon the olfactory bulb may provide central modulation of the sensory inflow. Thus, in the olfactory bulb, afferent fibers from the anterior olfactory nucleus synapse on the descending expansion (axon) of the internal granular cells. The presence of this axo-axonal synapse suggests that presynaptic inhibition (Eccles, 1961) may exist in the olfactory bulb.[2] The physiologic studies of Phillips, Powell, and Shepherd (1961), Yamamoto and Iwama (1962), Yamamoto, Yamamoto, and Iwama (1963), and Shepherd (1963) support the existence of an inhibitory action of the olfactory-bulb afferent fibers upon the dendrites, body, and axon of the internal granular cells and of the axons of the latter upon the horizontal dendrites of mitral cells.

Recent electron-microscopic studies of Hirata (1964) have shown the existence of "atypical" synapses in the olfactory bulb of the mouse. This author describes, among other synaptic figures, the presence of vesicular structures on both neuronal processes constituting synaptic junctions in the external plexiform and mitral-cell layers. He describes also synaptic junctions occurring on the surface of mitral cells with synaptic vesicles on the side of the neuron soma and in the apposed neuronal process, and also morphologically reciprocal or bipolar synaptic junctions in the external plexiform layer.

THE NUCLEUS OLFACTORIUS ANTERIOR AND
THE COMMISSURA ANTERIOR

The nucleus olfactorius anterior appears as an ovoid expansion partially lying within the caudal extent of the olfactory bulb. Its major axis runs rostrocaudally. This formation (Fig. 27, A.Olf.) is flanked laterally by the tractus olfactorius lateralis (T.Of.) and the fibrillar layer (F.L.) of the piriform lobe. Medially and dorsally it is covered by the stratum of commissural fibers (C.A.p.b.); ventrally and caudally it is continuous with the tuberculum olfactorium.

[2]Presumably similar are the contacts upon the amacrine cells of the retina described by Cajal (1893, 1911). The *serial synapses* on the amacrine trunks described by Gray and Young (1964) in the vertical lobe of the octopus are probably similar in nature.

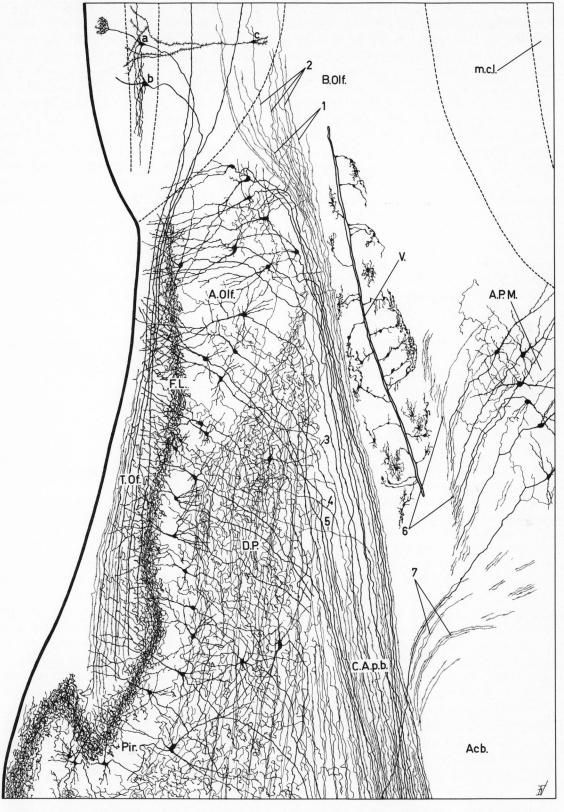

Fig. 27. Rat 43R, albino, 9 days old. Origin of the commissura anterior, pars bulbaris. Horizontal section through the nucleus olfactorius anterior (A.Olf.) and area praepiriformis (Pir.). Axons of pyramidal cells of the nucleus olfactorius anterior enter the pars bulbaris of the commissura anterior (C.A.p.b.); collateral branches of these axons enter the internal plexiform layer of the olfactory bulb (fibers labeled 2). Group of fibers labeled 1 are the axons of pyramidal cells of the contralateral nucleus olfactorius anterior. Golgi method.

It should be noted that the nucleus olfactorius anterior extends rostrally into the olfactory bulb, and therefore, just caudal to this, a line drawn transversely through the olfactory peduncle passes through the rostral portion of the nucleus olfactorius anterior.

The nucleus olfactorius anterior has been studied by a number of investigators (Herrick, 1924a; Gurdjian, 1925; Young, 1936; Crosby and Humphrey, 1939; Lauer, 1945; Lohman, 1963), who, on the basis of cytoarchitectonics, subdivided it. In Golgi material the cells of this nucleus appear more or less pyramidal in shape, with short dendrites radiating in all directions, sometimes resembling the basal dendrites of typical pyramidal cells, and one, two, or more long dendrites directed toward the fibrillar layer of the cortex piriformis (Fig. 27, F.L.) and the tractus olfactorius lateralis (T.Of.). These latter dendrites divide as they approach the dense plexus of the fibrillar layer. The cells each give off a thick axon which enters the commissura anterior, pars bulbaris (Fig. 27, C.A.p.b.). Before entering this commissure each axon emits retrograde collaterals to the nucleus olfactorius anterior and a branch, sometimes two, penetrating the internal plexiform layer of the homolateral olfactory bulb (group of fibers 2 in Fig. 27).

From Golgi material I gained the impression that in the rat the bulk of the fibers of the commissura anterior, pars bulbaris, originate largely in the nucleus olfactorius anterior; these fibers cross to the contralateral hemisphere, where some end in the nucleus olfactorius anterior and others continue farther to enter the internal plexiform layer of the olfactory bulb. This is illustrated in part in Fig. 27, where the group of fibers (1) entering the left olfactory bulb are the axons of neurons located in the right nucleus olfactorius anterior.

Therefore, the commissura anterior, pars bulbaris, is a commissural bundle which connects the nucleus olfactorius anterior of one side with its homologue of the other side, as well as with the contralateral olfactory bulb. There are few, if any, contributions from tufted cells of the olfactory bulb to the commissura. These cells, as mentioned before, are considered a variety of periglomerular cell and lack long projecting axons.

In the rat the commissura anterior, pars bulbaris, at its point of entrance into the olfactory bulb contains the axon collaterals of the cells of the homolateral nucleus olfactorius anterior (group of fibers 2 in Fig. 27) and the main axonal branches of the cells of the contralateral nucleus olfactorius anterior (group of fibers 1 in Fig. 27).

Figure 27 demonstrates that the circuit connecting both the olfactory bulb and the nucleus olfactorius anterior with their homologues of the contralateral side is as follows: impulses received by mitral cells a and b are transmitted through their axons to the fibrillar layer (F.L.), where synapses are made with the apical dendrites of cells of the nucleus olfactorius anterior (A.Olf.); the axons of these cells cross to the opposite hemisphere, via the commissura anterior, pars bulbaris, enter the contralateral nucleus olfactorius anterior, and synapse, also, with the internal granular cells of the contralateral olfactory bulb. Moreover, this circuit can drive the same mitral cell that initiated the circuit as follows (Fig. 27): axon of the mitral cell a–pyramidal cell of the homolateral nucleus olfactorius anterior (A.Olf.)–group 2 of fibers–internal granular cell c–horizontal dendrites of mitral cell a. The contralateral excitation obtained as explained previously would form the projecting part of the circuit; the reexcitation upon the same mitral cell that initiates the circuit, its feedback part.

The preceding description has been schematized in Fig. 28 in which the arrows 2 and 3 represent the tractus olfactorius lateralis transmitting impulses to the nucleus olfactorius anterior where impulses are relayed (a) through arrows 4, 7, and 8 to the contralateral olfactory bulb and nucleus olfactorius anterior via the commissura anterior, pars bulbaris, and (b) through arrow 5 to the homolateral olfactory bulb. Arrows 5 and 7 represent afferent olfactory-bulb impulses (conveyed by the group of fibers 2 and 1 in Fig. 27) to the internal granular cells, which, in turn, project through arrows 6 and 9 to the mitral cells.

If this is the case, the bilateral activation of the anterior parts of the piriform lobe and the contralateral activation of the olfactory bulb is originated in the same point (arrow 1 in Fig. 28) and the existence of two distinct pathways: a mi-

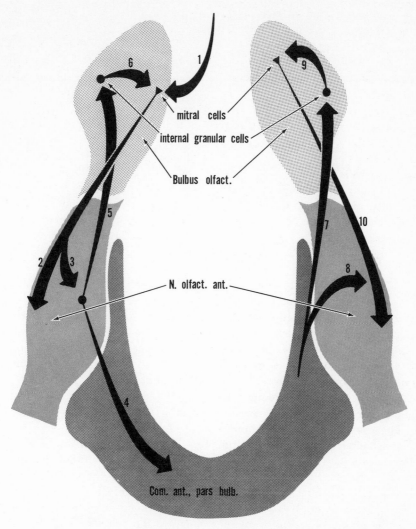

Fig. 28. Schema of interbulbar connections. Mitral cells of the left bulbus olfactorius send olfactory impulses (arrow 1) to the homolateral nucleus olfactorius anterior through arrows 2 and 3. From here impulses are transmitted through arrow 5 to the internal granular cells in the left bulbus olfactorius, or through arrows 4 and 7 to the internal granular cells in the right (contralateral) bulbus olfactorius. The internal granular cells, in the left bulbus olfactorius, close this circuit through arrow 6, and drive the mitral cells through arrow 9 in the right (contralateral) bulbus olfactorius.

tral cell–olfactory tract system, and a tufted cell–anterior commissure system, as proposed by Cajal (1911) and accepted by Allison (1953), can no longer be supported.

The afferent fibers to the olfactory bulb were first studied experimentally by Löwenthal (1897). Cajal (1890, 1901, 1911) recognized the existence of two classes of fibers entering the olfactory bulb:

thin axons of the contralateral olfactory-bulb tufted cells and thick ones, which he assumed proceeded either from the temporal lobe or from the olfactory peduncle. Cragg (1962) traced in the rabbit centrifugal fibers to the olfactory bulb after transection through the anterior cortex praepiriformis, tuberculum olfactorium, commissura anterior, and anterior neocortex. He found dense fiber degeneration around all the internal layers of granule cells and suggested that many of these fibers originate in the cortex praepiriformis.

Powell and Cowan (1963), after section of the lateral olfactory tract in the rat, traced degenerating fibers entering the homolateral olfactory bulb. They traced this system of centrifugal fibers to the granular cell layers, the external plexiform layer, and periglomerular cells. They suggested that this system of fibers arises in part at least from the olfactory tubercle. Their findings indicate the existence of centrifugal connections subserving a reflex control of activity in the olfactory bulb, and comparable in principle to the γ-efferent system in the spinal cord.

I believe that in the rat the two afferent olfactory-bulb fiber groups recognized by Cajal (1911) are the fine axon collaterals of cells in the homolateral nucleus olfactorius anterior (group *2* of Fig. 27) and the thick axons of neurons in the contralateral nucleus olfactorius anterior (group *1* of Fig. 27). In the present study of the rat no other types were observed entering the olfactory bulb, but this does not exclude the possibility of their existence. In this animal, as mentioned before, the axons of tufted cells do not leave the olfactory bulb. The assumption that the commissura anterior, pars bulbaris, arises largely from the nucleus olfactorius anterior is in accordance with the observations reported by Young (1936, 1941, 1942), Brodal (1948), and Orrego (1962).

With the Marchi technique Ban and Zyo (1962) traced in the rat fibers arising from the olfactory bulb, coursing through the commissura anterior, and terminating in the contralateral nucleus olfactorius anterior.

The existence of interbulbar fibers denied by Young (1941) is accepted by Probst (1901), Cajal (1911), Fox and Schmitz (1943), Clark and Meyer (1947), and Allison (1953), among others; however, recently Lohman and Lammers (1963) and Lohman (1963) reported that in guinea pigs the tractus olfactorius lateralis forms the only projection pathway of the olfactory bulb and that the pars rostralis of the nucleus olfactorius anterior projects, via the anterior limb of the anterior commissure, to the pars externa of the contralateral nucleus. These observations are in accordance

with the recent studies of Orrego (1962) in the turtle and the present studies in the rat.

The existence of interbulbar fibers became a subject of controversy because Cajal (1911) in his writings surmised that tufted cells project, via the commissura anterior pars bulbaris, to the contralateral olfactory bulb. Cajal (1911) carried out a series of experiments in the rabbit using the Marchi method in which, after ablation of one olfactory bulb, he traced degenerating fibers to the contralateral bulb. His remarks were erroneously interpreted by later investigators. Cajal made only a supposition: "Nous avons représenté sur la figure 423 un schéma de la marche *probable* des courants dans la voie principale constituée par les cellules mitrales et à houppette" (Cajal, 1911).

However, there are variations in the commissural pattern in different animals, as mentioned by Sonntag and Wollard (1925), Humphrey (1936), and Young (1936). This might explain the lack of interbulbar fibers in the rat (my Golgi observations) and guinea pig (Lohman and Lammers, 1963; Lohman, 1963) and their existence, for example, in the cat (Fox and Schmitz, 1943) and rabbit (Probst, 1901; Cajal, 1911; Clark and Meyer, 1947; Allison, 1953).

Recently the problem of the central modulation of the sensory inflow has attracted much attention. In connection with the olfactory system, several physiologic investigations on the influence of centrifugal fibers from the central nervous system to the olfactory bulb have been made (Arduini and Moruzzi, 1953; Kerr and Hagbarth, 1955; Angeleri and Carreras, 1956; Carreras and Angeleri, 1956; Kerr, 1960; Hernández-Peón, 1961; Yamamoto and Iwama, 1961; Mancia, Green, and Baumgarten, 1962). Anatomically, we possess evidence that centrifugal olfactory-bulb fibers synapse: (*a*) on the internal granular cells, where the possibility of presynaptic action on their descending expansions (axon) has been discussed (see p. 54), and (*b*) on the periglomerular cells (see p. 50) as described by Cragg (1962) and Powell and Cowan (1963).

Green and collaborators (Green, Mancia, and Baumgarten, 1962; Baumgarten, Green, and Mancia, 1962) described inhibition in the olfactory bulb of the rabbit after antidromic stimulation of both the tractus olfactorius lateralis and the commissura anterior, pars bulbaris. These investigators presented evidence of a recurrent inhibition which they believed to be transmitted by the recurrent collaterals of tufted and mitral cells. However, it should be noted that these investigators failed to produce antidromic activation of tufted cells. This does not invalidate the possibility that recurrent inhibition, resulting after stimulation of the commissura anterior, pars bulbaris, might be conducted through pathways other than by the axons of tufted cells.

Shepherd (1963) described in rabbits excitation of tufted and granule cells by orthodromic and antidromic volleys in the olfactory nerves and lateral olfactory tract respectively. The circuitous pathways described in the present work (see p. 56), relating the two olfactory bulbs by means of the intercalated pyramidal cell of the nucleus olfactorius anterior, would explain sufficiently the similarities of effect upon the induced activity in the olfactory bulb obtained from a section of the commissura anterior (Kerr and Hagbarth, 1955) and from ablation of the cortex praepiriformis (Carreras and Angeleri, 1956). The fact that the stimulation of the tractus olfactorius lateralis can reach certain olfactory regions of the brain and thence affect the contralateral olfactory bulb (Angeleri and Carreras, 1956; Carreras and Angeleri, 1956) can be clearly understood on the basis of the proposed circuit explained for Fig. 27 and schematized in Fig. 28.

The anterior limb (hitherto erroneously labeled pars bulbaris) of the commissura anterior connects other parts of the piriform lobe in addition to those just discussed. According to Young (1936), fibers from the anterior piriform area and from the capsula externa join the anterior limb of the commissura. Lauer (1945) also described fibers from the anterior piriform area entering that bundle, and Orrego (1962) points out that the piriform cortices of both sides are connected through crossed fibers. Ban and Zyo (1962) have reported the existence in the rat of fibers arising in the tuberculum olfactorium and capsula externa and terminating in the homolateral olfactory bulb and in the contralateral nucleus olfac-

torius anterior via the commissura anterior, pars bulbaris.

As shown in Fig. 27, several pyramidal-cell axons in the anterior part of the cortex piriformis (Pir.) caudal to the nucleus olfactorius anterior (A.Olf.) traverse the deep plexus of the cortex (D.P.) and turn caudally to enter the anterior limb of the commissura (C.A.p.b.). These axons could be followed in the next horizontal Golgi section, ventral to that drawn in Fig. 27; the majority of these fibers travel a short distance in the commissura and later surround the tuberculum olfactorium laterally and, joined by axons from adjacent piriform regions, collect in compact bundles just caudal to the tuberculum and enter the medial forebrain bundle (Valverde, 1963a).

However, several axons could be followed through the anterior limb of the commissura anterior; these may be commissural fibers projecting to the contralateral anterior piriform area. Fibers (labeled 6 and 7 in Fig. 27) of the area paraolfactoria medialis (A.P.M.) surround the ventricular cleft (V.) of the olfactory bulb caudally and travel a short distance in the anterior limb of the commissura anterior (C.A.p.b.) lateral to the nucleus accumbens septi (Acb.). I could not follow these axons farther, although they presumably enter basal regions of the brain.

Figure 29 illustrates the caudal extent of the Golgi section drawn in Fig. 27. Both represent part of a horizontal section through the entire left hemisphere of a 9-day-old albino rat. Figure 29 illustrates the plan of the commissura anterior, pars temporalis (C.A.p.t.). This temporal part originates largely in the temporal cortex, although moderate contributions are made by the piriform lobe. As the commissura approaches the cortex piriformis (Pir.) it spreads out, covering a zone which extends rostrocaudally from the region designated by an arrow to the caudal limits of the temporal cortex. That part of the commissura entering the depth of the cortex piriformis changes its direction abruptly and forms a sheet of fibers (the capsula externa) which enters the deep plexus (layer IVb) obliquely. Their terminal ramifications form in part the dense plexus found in the layer IV of the piriform and temporal cortices (IVa and IVb).

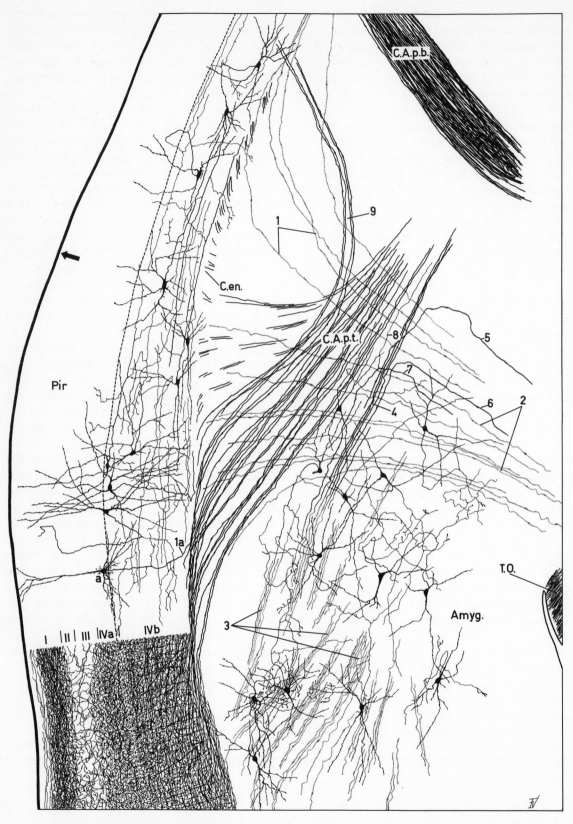

Fig. 29. Rat 43R, albino, 9 days old. Horizontal section through the cortex piriformis (Pir.) and amygdaloid complex (Amyg.), showing the fibrillar structure of the cortex, structural pattern of the pars temporalis of the commissura anterior (C.A.p.t.), and horizontal cells in the subdivision IVb of the cortex. Golgi method.

Conversely, the commissura anterior, pars temporalis, in its efferent component is formed by the axons (*1a*, Fig. 29) of pyramidal cells (cell *a*) connecting the piriform and temporal cortices with their homologues on the contralateral side.

As illustrated in Fig. 29, in subdivision IVb the small fascicles of fibers traversing this stratum perpendicularly predominate as well as fascicles more or less horizontally oriented; the former are axons of pyramidal cells entering either the commissura, pars temporalis (axon *1a*), or taking a more caudal exit through the amygdaloid region (group of fibers labeled *3*); the latter represent the final branches of bifurcation of fibers coming through the commissura, pars temporalis, from the contralateral side, as well as the axons and their collaterals of certain horizontal cells located throughout the entire rostrocaudal extent of layer IVb. I have drawn several cells of this type in Fig. 29. It can be observed that their axons and dendrites course in a horizontal direction, giving off several collaterals, also oriented horizontally. These cells predominate in the depth of the IVb subdivision and are a specialized cell type which interconnects distant parts of the cortex piriformis.

The superficial subdivision of layer IV (IVa) is formed by densely packed pericellular nests composed of fine fibers—the multiple subdivisions of the axon collaterals of large pyramidal cells—as well as other intrinsic fibers of the cortex.

As mentioned previously, some of the axons of cells in the caudal part of the piriform and temporal cortices do not enter the commissura, pars temporalis; these fibers form either the posterior limb of the capsula interna or part of the ansa lenticularis. The most ventral of these extracommissural fibers traverse the amygdaloid region (Fig. 29, group of fibers labeled *3*), and, coursing rostrally, join the commissura anterior, pars temporalis, as do, for example, the group of fibers labeled *4*. This is an aberrant bundle in which several fibers, such as those labeled *5, 6,* and *7,* turn sharply medially and leave the pars temporalis of the commissura anterior. Fiber *7*, at the point where it changes its course, gives off a collateral labeled *8*. The destination of these fibers (*5, 6, 7*) could not be determined. Bundle *9* (Fig. 29) leaves the temporal part of the commissura anterior, turns laterally, and enters the anterior

regions of the cortex piriformis. Small groups of fibers (see *1* and *2*) located ventral to the commissure project to caudal levels of the brain.

A piriform and a temporal component are recognized in the commissura anterior. The former comprises the part of the commissura that connects the nucleus olfactorius anterior, the anterior piriform area, and parts of the medial and posterior piriform areas with their homologues of the contralateral side, the two former through the anterior limb, the latter through the pars temporalis. The temporal component forms the rest of the commissura connecting both temporal cortices through the pars temporalis. Thus, there is a homology between the anterior limb (improperly termed the pars bulbaris) and the pars temporalis (often called the posterior limb); the anterior limb, excluding the link established between the nucleus olfactorius anterior and the contralateral olfactory bulb, connects cortical areas; the pars temporalis does the same.

In my Golgi studies it was impossible to decide how much of the anterior piriform area caudal to the nucleus olfactorius anterior contributes to the anterior limb. It was also difficult to establish the rostral border of the pars temporalis. This was rather arbitrarily designated at the point marked by an arrow in Fig. 29.

With these considerations, it can be assumed that in the commissura anterior the temporal component (not to be confused with the temporal limb) with respect to the piriform component and, likewise, the fasciculus longitudinalis associationis of the amygdala with respect to the stria terminalis, increase in number of fibers as they ascend in the phylogenetic scale. Both cases represent two examples of evolution in accordance with the studies of Johnston (1923), Bailey, Garol, and McCulloch (1941), Fox, Fisher, and Desalva (1948), and Klingler and Gloor (1960), among others.

THE FASCICULUS PROSENCEPHALI MEDIALIS (MEDIAL FOREBRAIN BUNDLE) AND RELATED STRUCTURES

Ventral to the commissura anterior is a complex region of the basal brain comprising numerous formations. This part of the brain (basal area

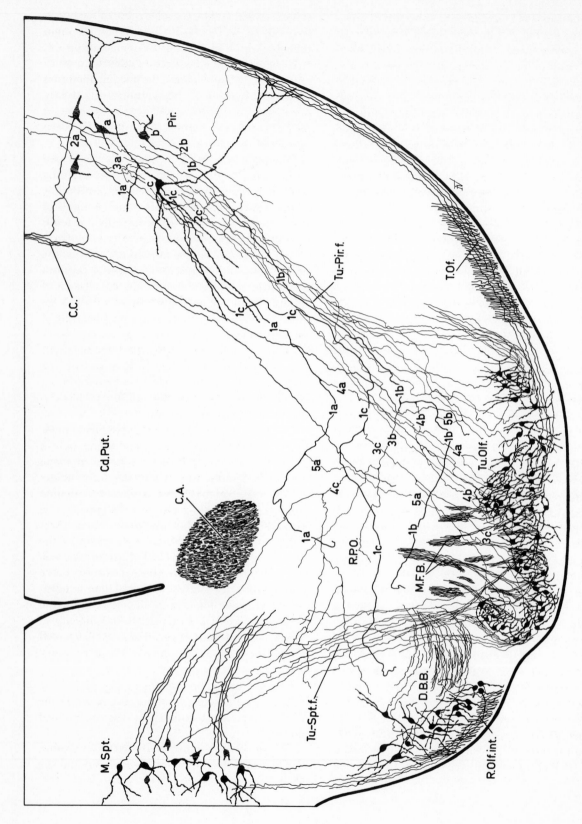

Fig. 30. Mouse 17, 3 days old. Transverse section through the tuberculum olfactorium (Tu.Olf.), showing the origin of one of the components of the medial forebrain bundle and the relations of the tuberculum with the septum and cortex piriformis. Golgi method. (From Valverde, 1963a, c.)

of Johnston, 1911) includes the tuberculum olfactorium (see Fig. 30) flanked laterally by the piriform cortex, medially by the nucleus of the diagonal band of Broca and the base of the septum, and dorsally by the regio praeoptica. Caudal to the level of the tuberculum olfactorium the regio praeoptica merges laterally with the area amygdaloidea anterior while caudally it is continuous with the hypothalamus. In a rostrocaudal direction a complex system of short as well as long neuron chains (the medial forebrain bundle) string together these basal regions of the brain with the brain stem.

Figure 30 is a transverse section of a 3-day-old mouse brain at the level of the tuberculum olfactorium (Tu.Olf.). Medial to it the radiatio olfactoria interna (R.Olf.int.) appears as a group of loosely scattered fascicles representing subpial fibers that course between the nucleus of the diagonal band and the medial septal regions. This fine sheet conveys fibers of the nucleus olfactorius anterior, precommissural hippocampus, medial septal regions, and medial parts of the tuberculum olfactorium (Gurdjian, 1925; Young, 1936; Lauer, 1945).

In Fig. 30 three projections have been observed from the tuberculum olfactorium. (a) The tuberculo-septal fibers (Tu.-Spt.f.), originating in the medial third of the tubercle, ascend dorsally forming scattered fascicles of fine fibers that enter the medial septal region (M.Spt.). (b) The tuberculopiriform fibers (Tu.-Pir.f.), originating in the lateral two-thirds of the tubercle, course upward and laterally, entering layer IV of the cortex piriformis; this system of fibers, made of fine axons originating from the clusters of little cells of the tubercle, is reciprocated by piriform-tubercular fibers to be dealt with below. (c) A third efferent system of fibers originates diffusely from the tuberculum olfactorium and, forming a series of fascicles, ascends toward the medial forebrain bundle area, where it enters as one of the major components of this bundle (M.F.B.).

As shown in Fig. 30, the axons *1a*, *1b*, and *1c* of the piriform cells *a*, *b*, and *c* could be followed coursing in the direction opposite to the tuberculopiriform system (Tu.-Pir.f.). In the regio praeoptica (R.P.O.), these thick axons give rise to several collaterals which enter the tuberculum olfactorium (Fig. 30, *4a*, *5a*, *3b*, *4b*, *5b*, *3c*, and *4c*); the main branch (*1a*, *1b*, and *1c* in the regio praeoptica, R.P.O.) turns caudally, ventral to the commissura anterior (C.A.) and enters the medial forebrain bundle. This system of fibers, comprising the tuberculo-piriform fibers as well as their opposites, is part of the rostral portion of the layer IV (deep plexus) of the piriform cortex.

The regio praeoptica forms an ill-defined area of the basal brain through which several long systems of fibers of the medial forebrain bundle traverse toward the hypothalamus. The limits of this region are poorly defined: rostrally it extends ventral to the commissura anterior, pars bulbaris, to the anterior olfactory formations, that is, the nucleus olfactorius anterior and rostral piriform cortex which lies lateral to it; the dorsal area of the tuberculum olfactorium merges with this region; medially it extends to the medial wall of the hemisphere, where it continues caudally without clear delimitations, with the hypothalamus. Laterally, at chiasmatic levels, it fuses with the area amygdaloidea anterior with which it forms a wide area extending from the midline to the deep plexus of the cortex piriformis.

As shown in Fig. 31, the cells of the regio praeoptica (R.P.O.) have axons that extend only a short distance. Their bodies are ovoid in shape and medium in size; their scanty dendrites radiate in all directions, but some tend to be oriented transversally (see, for example, cells labeled *g*, *h*, and *i*). The perpendicular disposition of these dendrites with respect to the long axons of the medial forebrain bundle (M.F.B.) traversing rostrocaudally the regio praeoptica is worthy of note and suggests the possibility of abundant axodendritic synapses. The axons of cells of the regio praeoptica are of fine caliber. It was difficult to follow their tortuous course. These axons give off several short collaterals which interlace profusely with each other.

The sagittal section drawn in Fig. 31 is more suitable for the study of other components of the medial forebrain bundle than those mentioned previously:

(a) Some thick axons of the corpus callosum (C.C.) proceeding from pyramidal cells of the

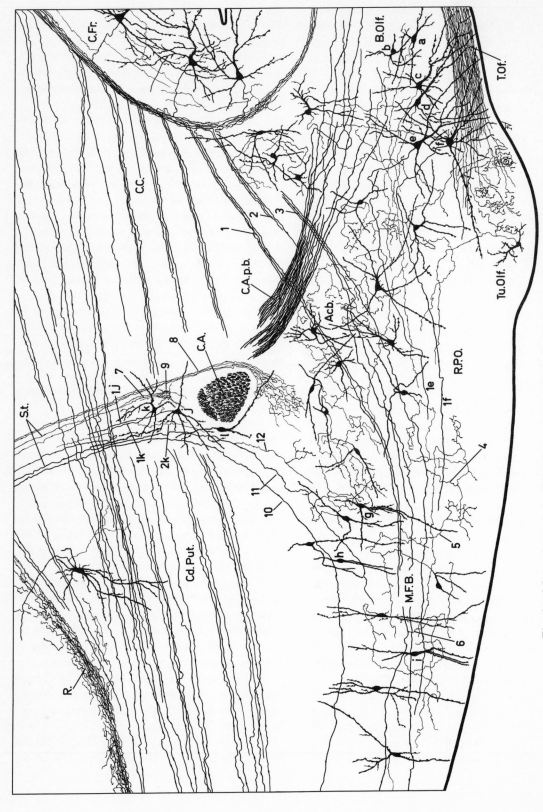

Fig. 31. Mouse 34, 6 days old. Sagittal section through the hypothalamus. Contributions of the stria terminalis and projection fibers of the frontal cortex to the medial forebrain bundle. Golgi method.

frontal cortex (C.Fr.), descend caudoventrally (bundles *1*, *2*, and *3*), course below the commissura anterior (C.A.), enter the regio praeoptica, and continue caudally as a component of the medial forebrain bundle (M.F.B.). These fibers give off collaterals (see fibrils labeled *4*, *5*, and *6*) that arborize among preoptic cells.

(*b*) Ventral to the commissura (Fig. 31, C.A.) fibers proceeding from the stria terminalis (*10*, *11*, and *12*, S.t.) enter the regio praeoptica, establishing contact with dendrites of cells in this region. The fiber labeled *10* could be traced farther caudally, but the majority of these fibers end a short distance ventral to the commissura.

(*c*) Axons of the septal cells *a*, *c*, *e*, and *h* (Fig. 32) take a descending course, cross through the commissura anterior to the regio praeoptica (R.P.O.) in which they turn in a sagittal direction, and enter the medial forebrain bundle. In the regio praeoptica, the axons give off some collaterals. In the same figure cells *f*, *g*, and *i*, belonging to the bed nucleus of the commissura anterior, project their axons in the same manner as do those of the septum just described. Cells *b* and *j* send their axons to the stria terminalis (its commissural component, S.t.com., is seen entering the commissura).

The majority of the axons of the cells of the septal nuclei form a condensed fiber system that enters the regio praeoptica and later courses in the medial forebrain bundle through the lateral hypothalamus. This system forms a large part of the Zuckerkandl's fascicle (Fig. 32, Z.R.).

To summarize, the medial forebrain bundle is composed of a long system of fibers and a short-axoned cell system. The first system includes the common pathway throughout the regio praeoptica and hypothalamus of the following five components:

(*a*) Fibers from the cortex piriformis. Axons of giant cells of layer IV course first medially to a point above the tuberculum olfactorium where they turn caudally in the medial forebrain bundle. This contribution is shown in Fig. 30 (axons *1a*, *1b*, and *1c*), Fig. 31 (cells *a* to *f* represent a portion of the rostral piriform cortex, the axons *1e* and *1f* from cells *e* and *f* course caudally in the medial forebrain bundle), and Fig. 33 (several axons of cells of the cortex piriformis lateral to the tuberculum olfactorium, Tu.Olf., can be traced to the medial forebrain bundle).

(*b*) Axons proceeding from the cluster of cells of the tuberculum olfactorium (M.F.B. in Fig. 30 and some axons of the cells of the tuberculum drawn in Fig. 33).

(*c*) Fibers of giant pyramidal cells of the frontal cortex located just above the olfactory bulb. These fibers course first among the fibers of the corpus callosum, later descend to the regio praeoptica, and then turn caudally below the commissura anterior (bundles *1*, *2*, and *3* of Fig. 31).

(*d*) The hypothalamic component of the stria terminalis (fibers *10*, *11*, and *12* of Fig. 31.

(*e*) Fibers originating from cells in the septal nuclei. The fibers descend just in front of the commissura anterior to the regio praeoptica, where they turn caudally.

The system of short links of the medial forebrain bundle is composed of cells extending from the regio praeoptica and lateral hypothalamus to mesencephalic levels. It includes all of the short-axoned cells of the regio praeoptica which, through their profusely interlaced short axons and collaterals, form the so-called *bed nucleus of the medial forebrain bundle.* They appear to be a caudal continuation of the other two bed nuclei of the forebrain, namely, the bed nucleus of the stria terminalis and the bed nucleus of the commissura anterior.

Rostral to the regio praeoptica the diagonal band of Broca forms a strip of cells and fibers extending transversally from the medial wall of the hemisphere to the cortex piriformis, with which its deep plexus becomes continuous. This disposition is shown in Fig. 33, which represents a transverse section through the tuberculum olfactorium in a 4-day-old albino rat. This section is slightly more rostral than Fig. 30. Figure 33 shows a transverse band delimited by two broken lines. This area corresponds to the region occupied by the diagonal band of Broca (D.B.B.). Medially in this area there is a group of cells of polymorphic type; cell *g* resembles a pyramidal neurone whose axon (*1g*) can be followed laterally, coursing between the medial forebrain bundle (M.F.B.) and the pars bulbaris of the commissura anterior

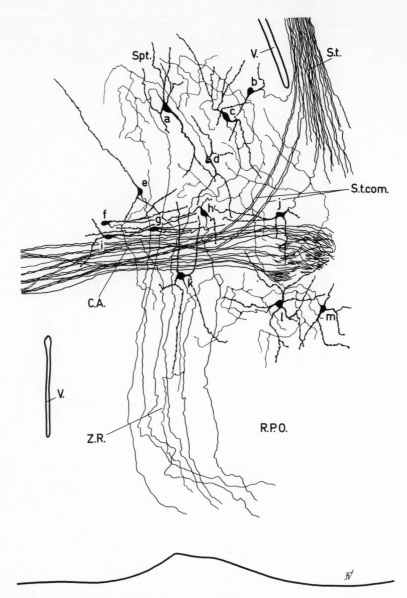

Fig. 32. Mouse 35, 6 days old. Transverse section through the commissura anterior, showing some cells of the lateral septal nucleus and neighboring gray matter. The origin of the Zuckerkandl's fascicle (Z.R.) and contributions of the stria terminalis to the commissura (S.t.com.) are shown. Some cells surrounding the commissura possess dendrites piercing that bundle. Golgi method. (From Valverde, 1963a.)

(C.A.p.b.), to a point where the diagonal band merges with the deep plexus (D.P.) of the cortex piriformis. The axons of the rest of the cells of this group (nucleus of the diagonal band) could be traced to the region occupied by fibers of the medial forebrain bundle. Cell *f* (a polymorphic cell of the deep plexus) sends its axon in the opposite direction to the axons of the cells just described.

In the same figure (Fig. 33) axons of cells of the cortex piriformis enter the deep plexus, where they emit several collaterals, giving off numerous

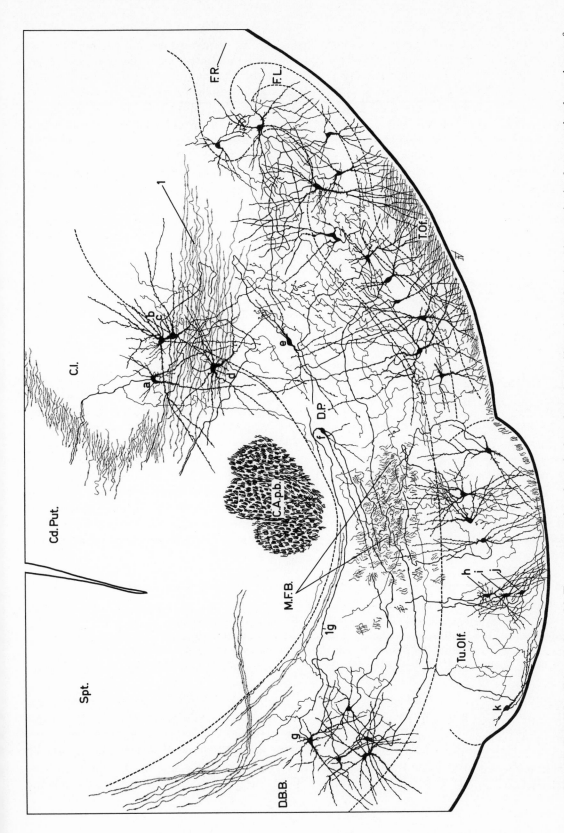

Fig. 33. Rat 41L, albino, 4 days old. Transverse section through the tuberculum olfactorium, showing the continuity between the deep plexus of the cortex piriformis (D.P.) and diagonal band of Broca (D.B.B.). Pyramidal cells of the rostral piriform regions with dendrites piercing the fibrillar layer (F.L.) and tractus olfactorius lateralis (T.Of.). Ascending axons of the tuberculum olfactorium (Tu.Olf.) enter the medial forebrain bundle (M.F.B.). Golgi method.

delicate subdivisions which contribute to the dense arborizations of this plexus. The main axonal branches of these piriform cells course through the diagonal band toward the fascicles of the medial forebrain bundle. Occasionally these thick axons were seen turning caudally to enter the medial forebrain bundle (just discussed and illustrated for Fig. 30). Ventral to the medial forebrain bundle (Fig. 33), in the tuberculum olfactorium (Tu.Olf.), a characteristic cluster of cells is illustrated (cells *h*, *i*, and *j*). The axons of these cells ascend to the medial forebrain bundle emitting recurrent collaterals to the cluster. Lateral to this cluster a group of large polymorphic cells send their axons, also, to the medial forebrain bundle.

These observations show that the diagonal band represents a two-way system of fibers connecting the rostral cortex piriformis with the medial paraolfactory centers and vice versa. Some fibers, however, coursing in this band, which originate in the cortex piriformis, bend caudally and enter the medial forebrain bundle. It is extremely difficult to determine whether or not axons of cells of the nucleus of the diagonal band enter the medial forebrain bundle.

Caudal to the diagonal band of Broca, the area amygdaloidea anterior connects the amygdaloid complex with the regio praeoptica. In lower mammals this relation is made by a short-axoned cell system; in higher forms a system of fibers develops, the fasciculus longitudinalis associationis, connecting the amygdala with the regio praeoptica, which obtains its greatest development in primates and man. In the mouse the area amygdaloidea anterior extends from the level of the caudal end of the tuberculum olfactorium to the anterior pole of the amygdaloid complex with which it gradually merges (Brodal, 1947). Laterally it fuses with the deep plexus of the piriform cortex, while rostrally and medially it appears to become continuous with the diagonal band of Broca and the regio praeoptica.

As illustrated in Fig. 34, in early stages of development some cells in the medial part of the amygdaloid matrix (cells *a*, *b*, *c*, and *d*) tend to direct their axons medially toward the hypothalamus (Hyp.). Probably this represents one of the first links established between the amygdalo-piri-

form region and the diencephalon. In the mouse (Fig. 35) the area amygdaloidea anterior forms a poorly delimited area whose cellular structure does not differ from that of the regio praeoptica. My Golgi observations reveal that the cells of this region are of medium size with long dendrites extending in all directions. The cell bodies are piriform, stellate, or fusiform (Fig. 35, *a* to *d*). The outstanding characteristic is the large number of collaterals given off by the same axon; the axon itself does not extend far from the cell body. Figure 35 is a drawing from a transverse section of a 20-day-old mouse; the rectangle within the section of the inset diagram shows the area represented in the figure. One cell (*a*) with its complete axon can be seen; the axon (*1a*) gives off a collateral (*2a*) to cell *b*, a second collateral (*3a*) turns back to the parent cell, and the collateral *4a* of the same axon also contributes to the innervation of cell *b*. The other collaterals of the axon of cell *a* contribute to the formation of the nests labeled *1* to *6* in which the cell bodies were not stained; a total of 8 nervous units would be influenced by the axon of cell *a*. The complete course of the axon of cell *c* was not observed; the nests *7* and *8* are formed by its collaterals. Cell *d* behaves in the same manner as cell *a*; it possesses an axon (*1d*) whose collaterals form the nests *9*, *10*, and *11*; the collateral *2d* turns back to the parent cell.

This arrangement can be considered the basic cellular organization of the area amygdaloidea anterior. In the mouse and rat I have not observed cells with long axons; if they exist, they must be rare.

Previous studies (Valverde, 1962, 1963a) suggested that the cells with short axons of the amygdaloid region form a linked system that extends through the amygdaloid nuclei and farther forward to the area amygdaloidea anterior and regio praeoptica. This system of short links was interpreted as the homologue of the ventral amygdalofugal pathway.

This complex part of the basal brain, comprising the several ill-defined and widely interconnected regions just described, emphasizes the multiplicity of connections of the olfactory centers. It is interesting to point out, however, that these centers also integrate emotional, behavioral, and in-

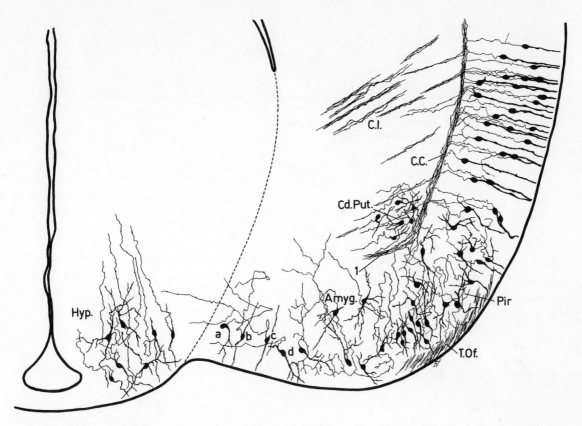

Fig. 34. Mouse 25, 14-mm embryo. Transverse section through the amygdaloid region. Note the globular shape of the immature cortical cells. Some amygdaloid cells (*a*, *b*, *c*, and *d*) send their axons medially toward the hypothalamus (Hyp.). Golgi method.

stinctive functions that arise from other distinct sources.

The common effector system of these basal regions of the brain is represented by the medial forebrain bundle and its short-linked system of cells, which I call the *medial forebrain bundle area*. In this area impulses coincide from regions such as the anterior olfactory formations, the cortex piriformis, the septal nuclei, the amygdaloid complex, the area amygdaloidea anterior, and the frontal cortex. In association with the hippocampal formation, interrupted by further relays, it projects to the midbrain tegmentum by way of the medial forebrain bundle, the stria medullaris–fasciculus retroflexus, and the mammilo-tegmental tract. These systems can be traced to the medial region of the midbrain to the area called by Nauta (1958) the *limbic midbrain region.* This circuit is reciprocated by ascending pathways toward the medial

forebrain bundle area, the hippocampus, and the amygdaloid nuclei. Ban and Zyo (1963) report the existence in the rabbit of ascending fibers, originated in the ventral tegmental nucleus of Gudden, the pars ventralis of the dorsal tegmental nucleus of Gudden and adjacent cells, and the lateral and ventral parts of the midbrain and pontile tegmentum, to the hypothalamus lateralis. Similar observations have been consigned by Cowan, Guillery, and Powell (1964), who found in the rabbit and the rat that the pars ventromedialis of the dorsal tegmental nucleus and the ventral tegmental nucleus contribute fibers, probably collaterals, to the medial forebrain bundle.

Guillery (1956), from the mammillary peduncle in the rat, and Nauta and Kuypers (1958), from the reticular formation of the pons in the cat, have traced degenerating fibers to the medial forebrain bundle and farther forward to the diagonal band

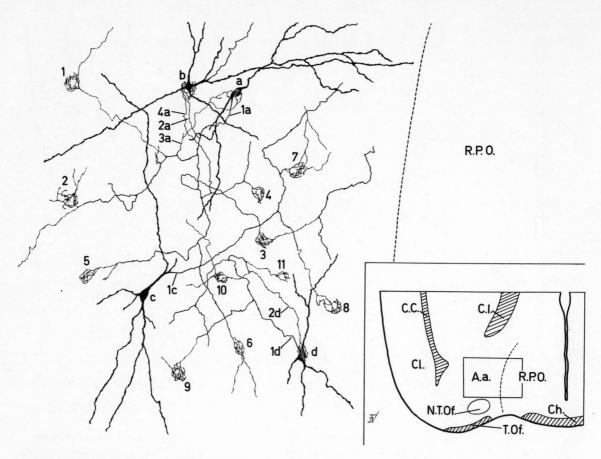

Fig. 35. Mouse 13, 20 days old. Transverse section through the area amygdaloidea anterior. The rectangle within the inset diagram shows the area drawn in the figure. Several cells in this area with multiramified axons, forming a short-linked chain of neurons, extend rostrally representing the substratum of the ventral amygdalofugal pathways in the mouse and rat. Golgi method. (From Valverde, 1963a.)

and medial septal nucleus. Morest (1961) has traced in the rabbit ascending fibers, from the dorsal tegmental nucleus, also to the medial forebrain bundle, nucleus of the diagonal band, and medial septal nucleus. Zyo, Ôki, and Ban (1963) described the existence in the rabbit of ascending fibers in the medial forebrain bundle to the septal nuclei. It is evident that such fibers represent the ascending part of the Zuckerkandl's radiation which was interrupted by the lesion of Cat 11 of my experimental series (see p. 29). From the septal nuclei impulses are relayed to the hippocampus via the fornix (Crosby, 1917; Rose and Woolsey, 1943; Morin, 1950; Daitz and Powell, 1954; Green and Adey, 1956; Cragg and Hamlyn, 1957; Votaw and Lauer, 1963; Powell, 1963). This substantiates the exist-

ence of a closed limbic-system–midbrain circuit as suggested by Nauta (1958).

In agreement with Auer (1953) and Guillery (1957), the observations reported above demonstrate the existence, in the basal brain of the mouse and rat, of a great number of pathways composed of chains of short neurons which predominate over the long fiber systems.

It appears that the first who traced the medial forebrain bundle was Ganser (1882), who labeled it *basales Längsbündel;* this author suggested that the bundle originates in a group of cells below the commissura anterior. He traced it caudally, parallel to the fornix, as far as the interpeduncular ganglion. Bischoff (1900) described a similar bundle which he called the *fasciculus olfacto-mesenceph-*

alicus; Wallenberg (1901) traced this bundle in the rabbit to the level of the trochlear nuclei and periaqueductal gray. Tsai (1925) pointed out that in the opossum part of the medial forebrain bundle does not bypass the hypothalamic region and considered the bundle as formed by several relatively short systems which included the tractus olfacto-hypothalamicus and the tractus olfacto-mammillaris; according to this author, long fibers reached the interpeduncular ganglion and tegmentum of the midbrain. Krieg (1932) also traced the medial forebrain bundle in the rat to the tegmentum. Loo (1931) described in the opossum the tractus olfacto-hypothalamicus which separates from the tractus olfactorius lateralis, pierces the tuberculum olfactorium, and joins the medial forebrain bundle.

Tello (1936) described three principal sources of origin of the medial forebrain bundle in the mouse: septal, frontal (in which several fibers appear to originate in the olfactory bulb), and temporal (which originates in its ventral region). These three groups of fibers pierce the hypothalamus in a rostral to caudal direction where many fibers are added from the region ventral to the striatum. Crosby and Woodburne (1951) consider that certain tracts extending caudally throughout the hypothalamus—the tracti hypothalamus-tegmentalis anterior and posterior—are short systems of chains connecting hypothalamic vegetative centers with the tegmentum and are distinct tracts from the medial forebrain bundle. Lohman and Lammers (1963), in the guinea pig, and Ban and Zyo (1962), in the rat, have traced fibers in the medial forebrain bundle originating in the nucleus olfactorius anterior.

Nauta (1956) traced degenerating fibers from the hippocampus via fornix to the regio praeoptica in the rat. Valenstein and Nauta (1959) pointed out that in the rat and guinea pig the medial forebrain bundle does not receive direct contributions from the hippocampus, but when lesions are placed in the septum, degeneration can be traced to the hypothalamus lateralis through the medial forebrain bundle. Similar observations have been described by Johnson (1957) in the mole. A reciprocal pathway from the regio praeoptica, relaying in the septum, to the hippocampus has been traced by Cragg (1961b) in the rabbit and rat. Nauta (1958) emphasizes that indirect hippocampal pathways, relayed in the preoptic and hypothalamic regions and septum, reach the midbrain through the medial forebrain bundle; however, in the cat (Valenstein and Nauta, 1959), direct hippocampal-fornix fibers reach the hypothalamus lateralis.

My observations on the composition of the medial forebrain bundle are in agreement with previous studies on this subject. The fibers described by Tello (1936) proceeding from the frontal and temporal cortices appear to be similar to those I observed coursing in the corpus callosum and tuberculo-piriform system.

Electrophysiologic observations (Ward and McCulloch, 1947; Sachs, Brendler, and Fulton, 1949) and anatomic studies (Bonin and Green, 1949; Clark and Meyer, 1949; Meyer, 1949; Wall, Glees, and Fulton, 1951; Beck, Meyer, and Lebeau, 1951; Nauta, 1962; Koikegami, 1963b) have shown the existence of direct connections between the orbitofrontal cortex and the medial forebrain bundle area.

The observation that the tuberculum olfactorium also contributes fibers to the medial forebrain bundle was reported early by Gurdjian (1925). This author also described further contributions from the paraolfactory area and septal region. These septal fibers have been considered as one of the major components of the medial forebrain bundle forming the so-called Zuckerkandl's fascicle. This fascicle was first described by Zuckerkandl (1888) as the *olfactory bundle of Ammon's horn;* he limited the account of this fascicle to the statement that its fibers descend very near to the midline in front of the commissura anterior to the anterior perforated substance, where it becomes intermingled with the numerous cells of the regio praeoptica. According to Cajal (1911), this fascicle contains descending and ascending fibers; the latter come from the brain stem. Concerning its descending fibers, Cajal denied their origin in the septum and pointed out that they arise in the gyrus fornicatus, pierce the corpus callosum, and, forming the fornix longus, descend through the septum to continue among the sagittally running fibers of the forebrain. According to my observations, the statements of Cajal are at least partially true, since, as reported previously (Valverde, 1963a) and as drawn in Fig. 32, I have observed cells of the septum sending their axons ventrally to reach the regio praeoptica and farther caudally. Cajal points out that the descending fibers of the septum course in the internal arciform fibers, which I believe are the same as the diagonal band of Broca. My observations have shown that the Zuckerkandl's fascicle is the direct route from the septum to the regio praeoptica and, in addition to septal contributions to the diagonal band, as reported previously, the bulk of the septal fibers descend around the commissura anterior, giving off many collaterals to the regio praeoptica and passing caudally as fibers coursing in the medial forebrain bundle.

The Zuckerkandl's fascicle also conveys afferent impulses to the septum, but these fibers are much less abundant than those coursing in the opposite direction; they seem to arise in the regio praeoptica. The septal fibers descending to the regio praeoptica continue caudally in the medial forebrain bundle, probably to reach mesencephalic levels. Nauta (1958), confirming his previous observations (Nauta 1956), traced such fibers, in the rat, to the ventral tegmental area.

The contributions of the stria terminalis to the medial forebrain bundle reported in the present observations, is represented by bundles 2 and 3 of Johnston (1923). These bundles descend behind the commissura anterior to the lateral hypothalamus, where they spread out ventrally, giving off numerous short collaterals. Adey and Meyer (1952) have shown that fibers running in the stria terminalis reach the infundibular region in the monkey. Kappers, Huber, and Crosby (1936) have traced stria terminalis fibers to the premammillary nuclei. Earlier workers, such as Sprenkel (1926), pointed out that bundle 2 originates from the gray matter behind the commissura anterior and suggested that such fibers could be traced caudally in the medial forebrain bundle. Johnston (1923) states that bundle 2, the hypothalamic bundle, descends very near the commissura anterior and bends backward in the medial forebrain bundle to enter the hypothalamus. Bundle 3 has been traced by the same author into the gray region medial to

the head of the nucleus caudatus but not as far as Sprenkel (1926) has traced it; however, Cajal (1911) traced such fibers to the subthalamic region and stated that this bundle increases as it descends caudally to establish relations with the nucleus called by him the *interstitial nucleus of the projection pathway of the temporal cortex.*

My observations are in accordance with the studies of Cajal, since bundles 2 and 3, indistinguishable from each other in the mouse and rat, can be followed to the lateral hypothalamus but not beyond the subthalamic region. The bundles form a complicated network in which are found the numerous cells with short axons of the regio praeoptica and hypothalamus lateralis. These cells, which have been named here the *bed nucleus of the medial forebrain bundle,* appear to form part of the *interstitial nucleus of the projection pathway of the temporal cortex* (Cajal).

Recently Guillery (1959) described certain pathways in the cat that provide a direct route from the medial forebrain bundle to the nucleus medialis dorsalis of the thalamus, and he expressed the idea that this nucleus behaves "as a mechanism that records the changes produced in the medial forebrain bundle by hypothalamic activity." This latter observation allows us to conclude that an important pathway leaves the medial forebrain bundle area to enter the thalamus. As reported by Guillery (1959), it appears that the first who interrupted this connection experimentally were Clark and Boggon (1933). Nauta (1958), after lesions in the lateral preoptic region of the cat, traced, via the inferior thalamic peduncle, degenerating fibers to the nucleus medialis dorsalis of the thalamus and Fortuyn, Hiddema, and Sanders-Woudstra (1960) found a related pathway in the rat. Similar observations have been reported by Cragg (1961a) and Zyo, Ôki, and Ban (1963) in the rabbit. As observed by Nauta and Whitlock (1954), this pathway appears to be reciprocated by direct connections between the nucleus medialis dorsalis of the thalamus and hypothalamus. This hypothalamo-thalamic pathway and its reciprocal appear to be reinforced by massive and also reciprocal connections existing between the amygdaloid complex and area amygdaloidea anterior from one side and the nucleus medialis dorsalis of the thalamus from the other side (Fox, 1949; Fortuyn, Hiddema, and Sanders-Woudstra, 1960; Nauta, 1961, 1962; Valverde, 1963b, c).

If we consider that the amygdala is related with the regio praeoptica either through the stria terminalis or through the fasciculus longitudinalis associationis (or its equivalent in the mouse and rat, that is, the short-linked system of neurons connecting the amygdala with the regio praeoptica through the area amygdaloidea anterior) and if we consider also that this connection is reciprocated by the preoptico-amygdaloid fibers coursing through the substantia innominata, described by Nauta (1958), by the ascending fibers traced by Zyo, Ôki, and Ban (1963) from the medial forebrain bundle to the area amygdaloidea anterior and by the amygdalopetal component of the longitudinal association bundle, originated in the hypothalamic and preoptic areas, described by Pow-

ell, Cowan, and Raisman (1963) in the rat, we will realize that the medial forebrain bundle area, the amygdaloid complex, and the nucleus medialis dorsalis of the thalamus are all three reciprocally and bidirectionally interconnected.

These reciprocal connections existing between the nucleus medialis dorsalis of the thalamus, the medial forebrain bundle area, and the amygdaloid nuclei including the area amygdaloidea anterior form three points of a closed subcortical circuit which, starting, for example, in the nucleus medialis dorsalis, may function following either the sequence of the arrows 2–3–1 or that of 6–4–5 of Fig. 36; and then this circuit would admit four more possibilities, two in both senses, initiated either in the medial forebrain bundle area (regio praeoptica + hypothalamus lateralis in Fig. 36) or in the amygdaloid complex. These three structures possess direct connections with the orbitofrontal and

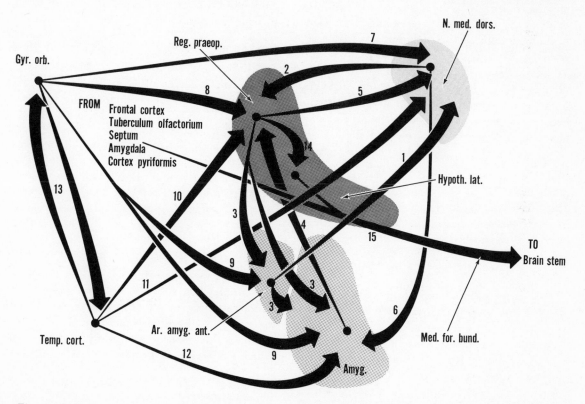

Fig. 36. Schema of the reciprocal connections between the amygdaloid complex–area amygdaloidea anterior, the nucleus medialis dorsalis of the thalamus, and the regio praeoptica–hypothalamus lateralis. These three regions are interconnected; they receive impulses from the orbital and temporal cortices. The medial forebrain bundle (arrow 15) represents a common effector pathway of these subcortical centers toward the brain stem.

temporal cortices, which lends support to the assumption that these cortices, themselves widely and bidirectionally interconnected (arrows 13) by the fasciculus uncinatus (Bucy and Klüver, 1940; Bailey *et al.*, 1943, 1944; Petr, Holden, and Jirout, 1949; Pribram, Lennox, and Dunsmore, 1950; Adey and Meyer, 1952; MacLean and Pribram, 1953; Pribram and MacLean, 1953; Whitlock and Nauta, 1956; Showers, 1958; Klingler and Gloor, 1960; Nauta, 1962), may exert direct and individual influences upon each of the three mentioned subcortical structures (centralization of Fessard, 1954) and thence upon the activity of the closed circuit interconnecting them (arrows 7, 8, and 9 from the gyrus orbitalis and 10, 11, and 12 from the temporal cortex in the schema of Fig. 36).

Connections between the orbitofrontal cortex and the medial forebrain bundle area have been mentioned (see p. 71); the existence of connections between the orbitofrontal cortex and the nucleus medialis dorsalis of the thalamus have been reported by Murphy and Gellhorn (1945), Rose and Woolsey (1948), Pribram and Bagshaw (1953), Nauta (1962), Koikegami (1963b), and Angevine, Locke, and Yakovlev (1964). Finally, connections between the gyrus orbitalis and the amygdaloid region were found in the course of the present work; see Cat 19, p. 43, and p. 116, paragraph (5). The relations existing between the temporal cortex and these structures will also be discussed later; see pp. 112–115.

THE CORTEX PIRIFORMIS AND CORTICO-AMYGDALOID RELATIONS

The cortex of the lobus piriformis has been subdivided into several regions based either on its cortical lamination or on its various anatomical connections. A detailed cytoarchitectonic analysis has been made by Rose (1929) but, of the several proposed subdivisions, that of Gray (1924) has been found to be the simplest and most convenient for my purpose. This author divides the cortex piriformis into the three areae piriformis anterior, medialis, and posterior, corresponding respectively to the area praepiriformis, periamygdaloid cortex, and area entorhinalis.

The area piriformis anterior corresponds to the *region of the olfactory tract* of Calleja (1893) and to the *écorce du lobe frontal sous-jacente à la racine externe* of Cajal (1911); the area piriformis medialis corresponds to the *sphenoidal cortex* of Calleja (1893) and presumably to the *région olfactive principale ou centrale de l'hippocampe* of Cajal (1911)[3] and the area piriformis posterior is the same as the *écorce temporal postérieure ou supérieure* of Cajal (1911) or area entorhinalis.

The area piriformis anterior of the mouse has been studied with the Golgi method by O'Leary (1937); the area piriformis posterior has been extensively studied by Lorente de Nó (1933), using also the Golgi method, and it will not be considered here.[4]

THE AREA PIRIFORMIS MEDIALIS OF THE CAT

I drew Fig. 37 directly from one original Golgi section of Cajal.[5] It represents a transverse section through the periamygdaloid cortex (area piriformis medialis) in a 25-day-old cat. In this section I find the following types of nerve cells:

(*a*) *Horizontal cells* (Fig. 37, *a*, *b*). These are small, stellate, or fusiform cells with dendrites coursing tangentially, some descending to the subjacent stratum. The axons could not be followed.

(*b*) *Small pyramids* (Fig. 37, *c*, *d*). These cells have many descending basilar dendrites and one apical dendritic branch, which divides, as soon as it leaves the cellular body, into several branches extending to the cortical limits. The axons of these cells descend to the deep layers, giving off several recurrent collaterals and collaterals directed tangentially.

(*c*) *Small polymorphic cells* (Fig. 37, *f*). These cells are about the same size as the horizontal cells in the first stratum, but they possess dendrites radiating in all directions, some of them ascending to the superficial stratum. The axon is a thick fiber descending toward the deep strata, which gives off several collaterals to the subjacent

---

[3]External, or olfactory, portion of the central cornual region (Cajal, 1955).

[4]For comparison of the nomenclature used in the lobus piriformis and related structures the reader is referred to Thomalske and Woringer (1957), Thomalske, Klingler, and Woringer (1957), and the Appendix in Gastaut and Lammers (1961).

[5]From the Cajal Museum at the Instituto Cajal, Madrid.

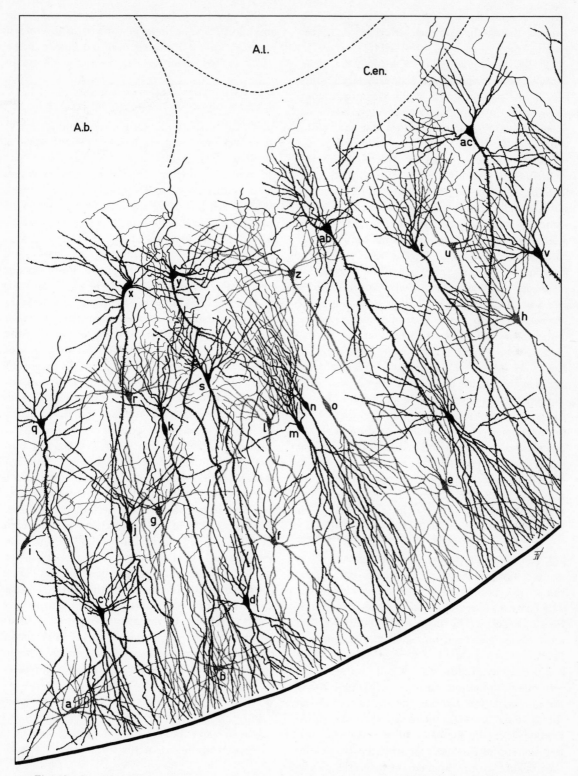

Fig. 37. Cat, 25 days old. Transverse section through the periamygdaloid cortex. A drawing made from one original Golgi section of Cajal.

layers. These cells have been found at the same level as the small pyramids described in paragraph (b).

(d) *Double-tufted cells* (Fig. 37, e, i, j, k, l, m, n, o, p). These cells have been found in large numbers in the middle stages of the periamygdaloid cortex; they are of medium size and spindle shape, possessing two bushes of dendrites, one projecting from each vertex. Of the two bushes, the one pointing inward is formed by thickly branched dendrites; the opposite one ascends to the superficial limits of the cortex and, although longer, is not generally so profusely ramified as the descending one. Occasionally one finds recurrent dendrites that reach the superficial limits of the cortex (see cell labeled l). The axons of these cells, arising generally from one of the descending dendrites far from the cell body (see cell o), descend to the deep stages of the cortex, giving off several collaterals to the layer of great pyramids, and finally enter the capsula externa.

(e) *Medium-sized pyramids* (Fig. 37, g, h, q, r, s, t, u, v, z). These are typical pyramidal cells located in the superficial limit of the deepest third of the cortex. They have a moderate number of basilar dendrites and one slightly ramified apical branch that ascends to the first stratum. The axons possess numerous collaterals, some of them following recurrent and ascending courses; the main axonal branch enters the underlying white substance and either courses in the capsula externa or pierces the amygdaloid complex between the nucleus lateralis and the nucleus basalis.

(f) *Deep great pyramids* (Fig. 37, ab, ac). These are giant typical pyramidal cells, each with a profuse basal dendritic apparatus and an apical branch ascending toward the superficial levels. The latter divides and gives off a moderate number of branches. The axons enter the underlying white substance, emitting several recurrent collaterals.

(g) *Deep polymorphic cells* (Fig. 37, x, y). These cells vaguely resemble giant pyramidal ones; they are located in the depth of the cortex and are of the same size as the giant pyramids; their apical dendritic branch, slightly ramified, does not follow the straight course taken by the pyramidal ones. The axon, directed toward the white substance, emits numerous recurrent collaterals

which ramify profusely in the overlying layers among the basilar dendrites of the medium-sized pyramids and descending terminal bushes of the double-tufted cells.

THE AREAE PIRIFORMIS ANTERIOR AND MEDIALIS OF THE MOUSE AND THE RAT

In the anterior piriform area Calleja (1893) described five layers as follows: (I) the fibrillar layer or layer of the lateral olfactory tract, (II) the molecular layer, (III) the layer of medium-sized pyramidal cells, (IV) the layer of giant pyramidal cells, and (V) the plexiform layer. With minor variants this division was followed by Cajal (1911) and O'Leary (1937).

Caudal to the termination of the direct olfactory fibers, the cortex piriformis shows certain structural differences with respect to the anterior piriform area. This part of the piriform lobe, as mentioned before, has been called the *sphenoidal cortex* by Calleja (1893) and corresponds to the *région olfactive principale ou centrale de l'hippocampe* of Cajal (1911).[6] In this region (also area piriformis medialis or periamygdaloid cortex), both authors, like Kölliker (1896), stress certain peculiarities such as the existence of the double-tufted cells, the giant polymorphic cells of the second layer, and the alternative existence in this layer of clusters of cells of giant polymorphic and little pyramidal cells. I have not observed these clusters. The existence of the double-tufted cells appears as a common feature in the periamygdaloid cortex; their existence was simultaneously discovered by Calleja (1893) and Kölliker (1896). These authors stressed that double-tufted cells are a distinctive characteristic of this part of the cortex piriformis. It should be noted that these cells are strikingly similar to certain cells described by Cajal (1911) in the insula.

The description given by Calleja (1893) under the name *sphenoidal cortex* to the periamygdaloid region in the rabbit, while similar in many points, differs somewhat with respect to Cajal's and my descriptions; presumably we have been studying slightly different zones, where minor variants might be expected.

In the mouse, as in the rat, both areae piri-

[6]See footnote 3.

formis anterior and medialis are very similar to each other; hence, the following description applies to both. Where relevant, the minor variants existing between both areae will be mentioned. In these animals the main feature of the cortex piriformis is the appearance under low magnification of two dense plexuses. The first plexus occupies the superficial stratum and extends throughout the entire cortex piriformis; rostrally it is formed by the axons of mitral cells and several kinds of fibers to be discussed below; caudally it continues to the subiculum. The second plexus occupies the deep half of the cortex and it is also present throughout the whole extension of the piriform lobe.

The appearance of these two plexuses can be seen in Fig. 29. It should be observed that in the layers II and III there are a small number of fibers among which densely packed cell bodies are located. The intrinsic organization of these two plexuses will be discussed later (see pp. 82–86).

According to Lorente de Nó (1933), fibrillar plexuses can be considered "*as regions of the cerebral cortex where special contacts take place.*" Undoubtedly, a given cortical layer is most easily identified by a sharply delimited cortical plexus, but between plexuses the stratification of the cortex must be based on the existence of neurons of similar appearance. This procedure is not appropriate, since there are often in a single layer several different kinds of cells which cannot be grouped under a predominant type. However, an attempt at stratification has been made based on the presence of small pyramidal cells in the superficial half of the cortical gray located between the two piriform plexuses and of medium-sized pyramids close to the superficial limits of the second (deep) plexus. The stratification of both areae piriformis anterior and medialis can be considered as follows:

First layer	Fibrillar layer
Second layer	Layer of small pyramidal cells
Third layer	Layer of medium-sized pyramidal cells
Fourth layer	Deep plexus

It should be pointed out that the names given to the second and third layers do not imply that the pyramidal cells are the most abundant type in these layers.

If my interpretations are correct, the subdivisions I_a and I_b given by O'Leary (1937) would correspond approximately to the fibrillar layer of the present study; layer II of O'Leary would correspond to my layers II and III, while layers III and IV of this author would be the same as the deep plaxus (D.P.) of the present study. Following the description given by Calleja (1893) and Cajal (1911), the stratification made by O'Leary is based mainly on the cell shape and dendritic and axonal patterns. In the present investigation the existence of two plexuses (fibrillar layer and deep plexus) has been taken as a principal objective criterion for the description of this part of the cortex.

Fibrillar layer

As this layer is almost exclusively occupied by fibers and dendritic terminations, it is difficult to observe any cells and to study their morphology. Its fibrillar structure will be considered below (see pp. 82–86).

Layer of small pyramids

The following types of cells are present in this layer:

(*a*) *Small pyramids* (mouse: Fig. 38, *c*; rat: Fig. 40, *r*; see also Fig. 33). These cells show relatively short basilar dendrites, the apical branch ascends toward the fibrillar layer, and, as shown in Fig. 40, *r*, it divides promptly into secondary branches which reach the first layer. The axon descends to the deep plexus without emitting any collaterals in the third layer. As shown in Fig. 38, the axon *1c* of the small pyramid labeled *c* emits several horizontal collaterals (*2c* to *5c*) occupying the superficial half of the deep plexus. The same occurs with respect to the axon *1r* of cell *r* of Fig. 40. The main branch of the axon courses through the deep plexus either to follow the path of the diagonal band of Broca (as can be observed for several of the axons of small pyramids in Fig. 33), or to enter other subcortical formations. The axon *1r* of cell *r* in Fig. 40 appears to ramify completely in the deep plexus (D.P.).

(*b*) *Semilunar cells* (mouse: Fig. 38, *b*, *d*). These

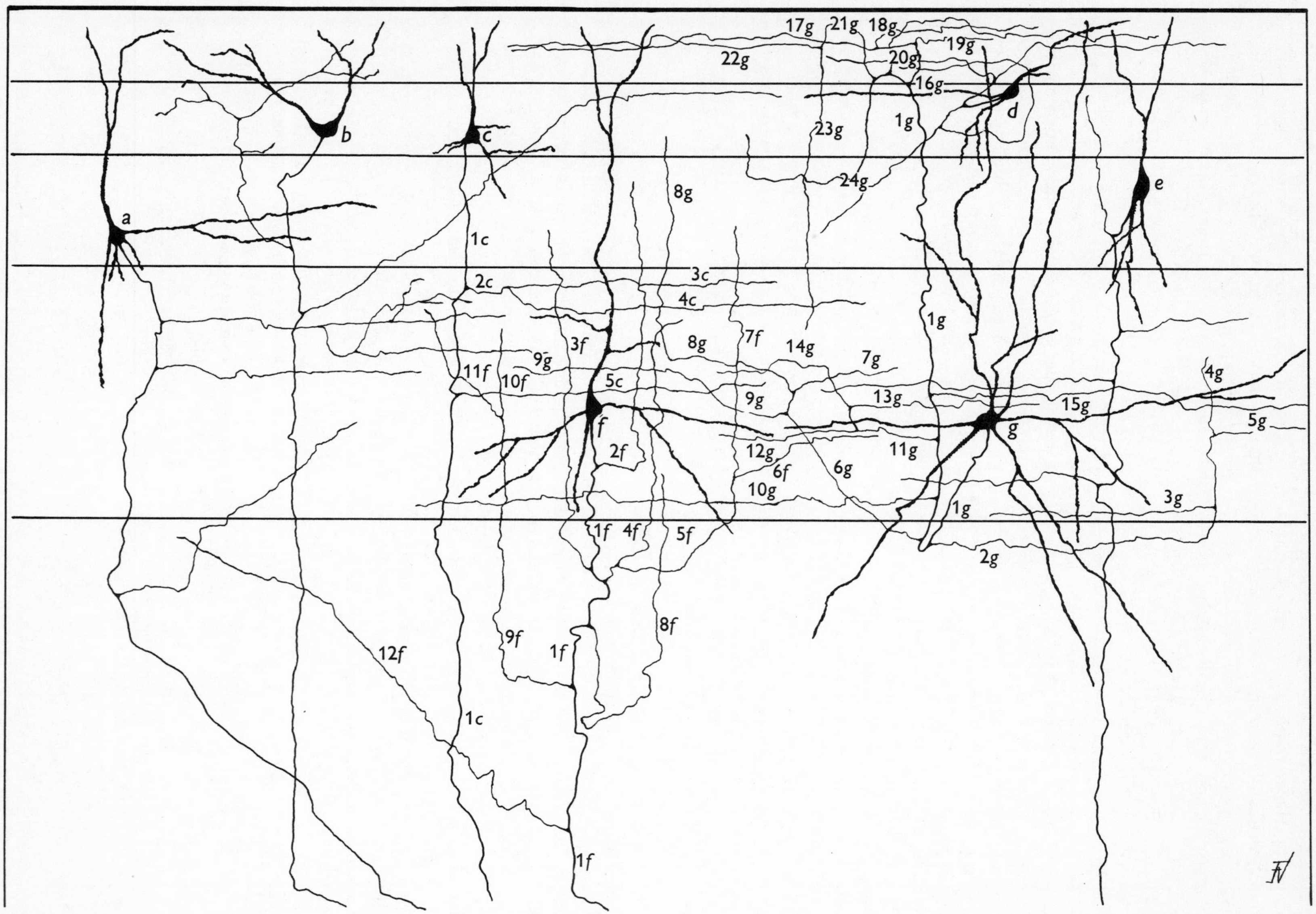

Fig. 38. Mouse 17, 3 days old. Piriform cortex. Cells of layers II, III, and IV with nearly complete axonal traject. Golgi method. (From Valverde, 1963a.)

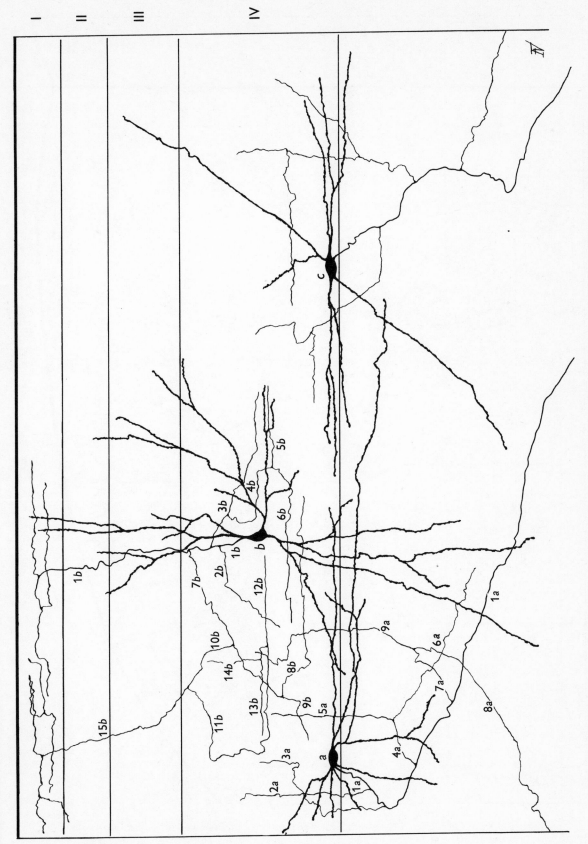

Fig. 39. Mouse 17, 3 days old. Piriform cortex. Stellate and horizontal cells in layer IV. Golgi method. (From Valverde, 1963a.)

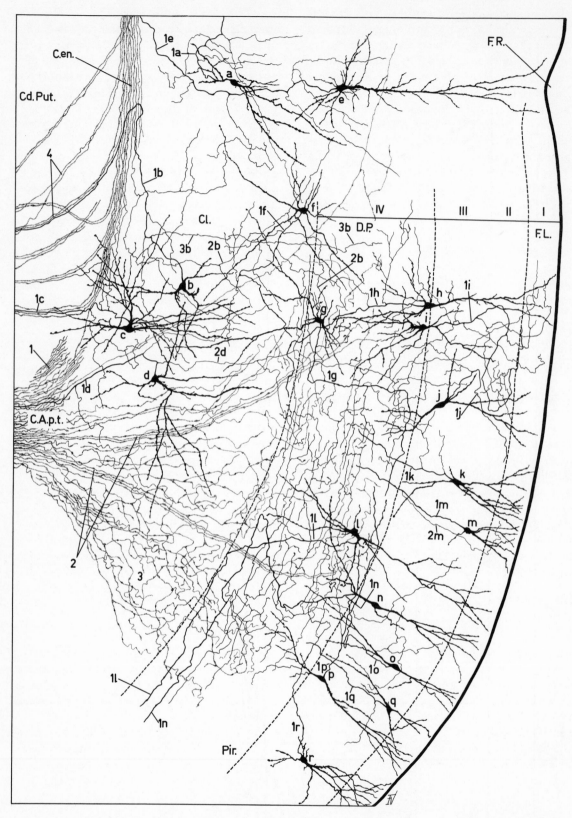

Fig. 40. Rat 41L, albino, 4 days old. Transverse section through the cortex piriformis. Some cells of layers II, III, and IV (*f* to *r*) and of the deep claustral gray (*a* to *d*) are shown. Ascending dendrites of piriform cells reach layer I, or the fibrillar layer (F.L.). Many axons of these cells course in layer IV, or the deep plexus (D.P.). Bundles (*2*) of the pars temporalis of the commissura anterior (C.A.p.t.) enter the deep plexus. Golgi method.

cells have been found frequently in the cortex piriformis; they emit dendrites in groups at the pole of their semicircular or globular bodies which ascend to the fibrillar layer. The axons project quite differently; the axon of cell *b* in Fig. 38 descends perpendicularly through the cortex piriformis to enter the underlying white substance; in the fourth layer it emits a retrograde collateral which ascends obliquely to the second layer where it courses tangentially. The axon of cell *d* of the same figure emits several collaterals close to its cell body, which course in the first layer.

(*c*) *Fusiform cells* (rat: Fig. 40, *o*). Few are present in this layer. They will be discussed in layer III.

(*d*) *Polygonal cells* (rat: Fig. 40, *k, m, q*). Cells varying in shape between triangular, globular, polygonal, and completely irregular are included under this term. A globular cell shape generally represents an immature stage of development. The process of dendrite growth changes the globular form of cells into a polygonal appearance (compare cells of Fig. 34 with those of Fig. 40). The use of young animals in the Golgi method causes the erroneous interpretation of cell shapes that actually represent transitional forms.

The axons of these cells descend to the fourth layer, giving off a short number of collaterals, either recurrent or more or less tangentially oriented, without definite predominance for a given layer. In the cat I observed polygonal cells (Fig. 45, *i*) with a complete axonal apparatus whose main branch coursed horizontally and distributed abundant collaterals through the second and third layers; some of them entered the superficial part of the fourth layer.

Layer of medium-sized pyramids

The following cells were identified in this layer:

(*a*) *Medium sized pyramids* (mouse: Fig. 38, *a, e*; rat: Fig. 29, *a*, Fig. 27, Fig 46, *j*). Sometimes the shape of these cells does not correspond to that of a typical pyramidal cell, as illustrated for cells *a* and *e* in Fig. 38. The apical dendritic branch of each of these cells subdivides near the cell body into several branches which extend to the fibrillar layer (Fig. 27, F.L.). When the bodies of these cells

lie in the depth of the layer, their basilar dendrites enter the deep plexus predominantly (Fig. 46, *j*). The axons of these pyramids pierce perpendicularly the underlying fourth layer, emitting horizontal collaterals (Fig. 38, *a*) which ramify profusely in this layer (Fig. 46, collateral *2j*). In general, the axons of these cells project separately in different patterns; in the rostral part of the cortex piriformis they may enter as fibers of the commissura anterior, pars bulbaris (Fig. 27, C.A.p.b.) or may enter the medial forebrain bundle (Fig. 31, axons *1e* and *1f* of cells *e* and *f*); caudally the axons become incorporated into the capsula externa (Fig. 46, C.en., group of fibers labeled *17* and axon *1j* of cell *j*), where they may emit collaterals to the nucleus amygdaloideus lateralis (Fig. 46, A.l., collaterals *3j* and *4j*), or they may cross to the contralateral piriform lobe through the commissura anterior, pars temporalis (C.A.p.t.), as illustrated with axon *1a* of cell *a* in Fig. 29.

(*b*) *Fusiform cells* (rat: Fig. 40, *j, n, p*). Their form closely resembles that of double-tufted cells of the cat's cortex piriformis, although their dendritic bushes are not as profusely ramified and are often represented by a single thick dendrite containing scanty short secondary branches. The peripheral group of dendrites reaches the fibrillar layer, whereas the descending group enters the deep plexus. The axons of these cells apparently enter and ramify only in the deep plexus.

The axon *1n* of cell *n* (Fig. 40) extends to the region of the tuberculum olfactorium; during its course through the deep plexus it gives off numerous horizontal collaterals. The axon *1p* of cell *p* in the same figure behaves in a similar manner, while the axon *1j* of cell *j* is an ascending axon emitting a descending collateral that reaches the fourth layer.

Deep plexus

The deep plexus is one of the most interesting structures of the piriform lobe. Its intrinsic organization will be discussed below (see pp. 82–86); the several kinds of cells present in this plexus will be described first:

(*a*) *Great pyramids* (mouse: Fig. 38, *f*, Fig. 39, *b*; rat: Fig. 40, *h*). The shape of these cells varies considerably; cell *h* of Fig. 40 appears similar to that of the classical pyramids, as does also cell *f* of Fig. 38, while cell *b* of Fig. 39 is quite different. In the latter, the axon is ascending, whereas in the former the axons descend.

(*b*) *Horizontal cells* (mouse: Fig. 39, *a*, *c*; rat: Fig. 29, all of the cells drawn in the depth of the IVb subdivision, Fig. 33, *e*, *f*, Fig. 42). These cells have been found particularly abundantly in this layer, especially in its deepest part. Their dendrites, like their axons and collaterals, show a tendency to course rostrocaudally. Thus, the axons of these cells form a system of fibers linking distant parts of the cortex rostrocaudally.

(*c*) *Polygonal cells with ascending dendrites to the first layer.* These cells correspond to the types reproduced in Fig. 30, *c*, Fig. 38, *g*, and Fig. 43*B* for the mouse and Fig. 40, *l* for the rat.

(*d*) *Polygonal cells with dendrites radiating in all directions.* These cells are reproduced in Figs. 41 and 43*A*, for the mouse and in Fig. 40, *i*, *g*, for the rat.

It should be added that the last two groups of cells could be classified also according to their axonal pattern of projection into at least four categories:

(1) Cells with ascending axons ramifying in the four layers (Fig. 40, *i*, and Fig. 43*A*).

(2) Cells with ascending axons emitting horizontal collaterals in the fourth and first layers (Fig. 38, *g*, and Fig. 41).

(3) Cells with axons coursing through the fourth layer, either to enter other subcortical structures or to ramify completely within this deep plexus (Fig. 40, *g*, *l*, and Fig. 43, cell drawn in *B* whose axon, labeled *12*, could be followed to the tuberculum olfactorium).

(4) Cells with axons coursing through the fourth layer, emitting several collaterals, and possessing ascending collaterals that ramify in the first layer. The red axon labeled *9* in Fig. 43*B* and its collaterals (the ascending ones have been labeled *10* and *11*) proceed from a cell (not drawn) similar to that of *A* of the same figure. These axons, although not represented, could be followed to the tuberculum olfactorium.

O'Leary (1937) describes in his layer IV a variety of fusiform cells provided by few lateral dendrites and two thick ascending dendrites frequently reaching the plexiform layer. He could not stain the axons of these cells, but he assumed that "they belong to the output mechanism of the cortex." According to his drawings this type of cell would rather correspond to cell *b* of Fig. 39 and the cell of Fig. 43*A*. I have observed that cells of this type possess ascending axons.

THE CONDUCTION OF IMPULSES THROUGH THE CORTEX PIRIFORMIS

The following attempt to ascertain the significance of the two plexuses of the cortex piriformis is based mainly on my Golgi observations in mouse and rat brains. Reference will be made also to the fibrillar layer of the cat's piriform cortex, even though a simple schema of the cortical organization in this area has been recogized in the former brains.

An essential difference exists between the two plexuses: in the fibrillar layer there are present almost exclusively axo-dendritic synapses, while in the deep plexus axo-somatic as well as axo-dendritic synapses occur.

At rostral piriform levels (see Fig. 27) the axons of mitral cells emit short collaterals, observed by Cajal (1891) and Calleja (1893), which collect, forming a layer of profusely interlaced thin fibrils (F.L.) along the medial aspect of the tractus olfactorius lateralis. These observations were confirmed by O'Leary (1937), who also states that olfactory tract axons terminate in his field I_b, entirely above the layer of superficial pyramids.

Innumerable ascending dendrites of cells from the underlying layers enter this plexus. The extent to which the piriform lobe is supplied by these mitral axons and their short collaterals could not be determined. As mentioned previously, in a thick horizontal section of the rat's brain it was possible to trace mitral axons to the periamygdaloid cortex. Mitral axons to the nucleus tractus olfactorii lateralis in the rat could be followed in sagittal series, but few, if any, mitral axons could be traced caudal to rostral amygdaloid levels. However, caudal to the region where mitral axons terminate, the plexus continues without apparent modification in its gross aspect (see layer I in Fig. 29). Here the fibrillar layer is formed by the final ramifications of ascending axons from cells in the underlying layers.

In the cat I observed different types of axo-somatic and axo-dendritic contacts in the fibrillar layer, and also axo-somatic contacts in the layer below. Figure 44 is a transverse section through the periamygdaloid cortex of a cat embryo in which fibers of the fibrillar layer as well as several of the most superficial cells have been reproduced. Through the fibrillar layer course fibers of various calibers; the fine ones emit at intervals small sprouts (labeled *8*, *9*, *10*, *11*, and *12*) that make synaptic contacts with ascending dendrites from the underlying cells. Below the fibrillar layer these axo-dendritic contacts are rarely observed (see point labeled *7*). The fiber labeled *3* in this layer climbs for a short distance one of the dendrites of cell *d*. In this figure several axo-somatic contacts have been reproduced; for example, fiber *2* synapses with the cell body *a*, fiber *3* terminates in the pericellular nest *4*, and the collaterals *2d* and *2g* of the axons of cells *d* and *g* also terminate within nests *6* and *5* respectively.

At rostral levels, where the fibrillar layer is present together with the most superficial stratum of mitral axons, the dendrites bypass the fibrillar layer and enter into the tractus olfactorius lateralis. This can be observed in Figs. 27 and 33. Whether or not these dendrites establish synaptic contact with the main axonal branch of the mitral cells (T.Of.) could not be determined; it seems that the site of synaptic contact must be located in the subjacent fibrillar layer (Figs. 27 and 33, F.L.). Caudal to the tuberculum olfactorium the tractus olfactorius spreads out and the fibrillar layer gains the surface; this is clearly shown in Fig. 27.

As mentioned previously, besides the collaterals furnished by the main axonal branches of mitral cells, the fibrillar layer is formed by short ramifications of ascending collaterals given off by some of the several kinds of cells described above. In Fig. 38 the axon *1g* of cell *g*, after giving off the collaterals *2g* to *15g* for the fourth layer, ascends without emitting collaterals through the third and second layers to the first layer where it emits the collaterals *16g* to *22g*. The ascending branches *1b* and *15b* of cell *b* in Fig. 39 behave in a similar manner. Another example is given in Fig. 41; the ascending branches *2*, *3*, and *4* of the cell drawn in Fig.

43*A* and the ascending collaterals *10* and *11* of axon *9* in Fig. 43*B* also provide additional examples of fibers contributing to form the fibrillar layer of the cortex piriformis.

The fibers of the first layer synapse with the ascending dendritic branches of the underlying cells; however, although rarely observed, some fibers of the fibrillar layer descend to establish contacts in the subjacent layers, as observed for fibers *5* and *6* of Fig. 43*A*.

The deep plexus or fourth layer is a complicated mazework, frequently confused with certain adjacent structures. At its rostralmost extremity it fuses with the commissura anterior, pars bulbaris (Fig. 27, D.P. and C.A.p.b.). Below the level represented in this figure it merges dorsally with the tuberculum olfactorium, including the medial forebrain bundle area, where it becomes continuous with the diagonal band of Broca (Fig. 33, D.P. and D.B.B.) and the tuberculo-piriform system of fibers (Fig. 30, Tu.-Pir.f.); caudally it includes part of the area amygdaloidea anterior and at its most caudal extent it includes fibers of the commissura anterior, pars temporalis (Fig. 29, C.A.p.t).

In frontal Golgi sections of the rat's brain I observed caudal to the area amygdaloidea anterior, where the amygdaloid complex gains the surface of the brain (nucleus amygdaloideus corticalis), that the deep plexus becomes in some places slightly displaced laterally and fuses partly with the fibrillar layer (Fig. 43*C*). At this level, the deep plexus also fuses partly with the capsula externa (C.en. in Fig. 43*C*), although this fusion is partly formed by a group of axons of the cortex approaching the capsula (Fig. 46, group of fibers labeled *17*).

Through the deep plexus course the axons of many cells of the piriform cortex, which give off a variable number of collaterals contributing to the mazework of the plexus. Included also in this deep plexus are the projection fibers of the piriform cortex, as well as corticopetal and intrinsic fibers, the latter corresponding to intracortical association axons and collaterals of long projecting ones.

On the basis on the preceding Golgi observations, part of the activity of the cortex piriformis has been schematized in Fig. 43*D*. This schema includes the olfactory inflow passing toward the medial forebrain bundle area, including the cortex piriformis as an intermediate link. Fibers of the tractus olfactorius lateralis (Fig. 43D, T.Of.) establish synaptic contact, through the fibrillar layer (F.L.), with apical dendritic branches of the piriform cells a, b, and c; from the fibrillar layer also, fibers could be traced to establish contact with these cellular bodies (see the blue fiber that, from the fibrillar layer, descends to contact with cell c, representing the fibers 5 and 6 of A of the same figure). A second olfactory inflow has been reproduced by the red fiber that descends from the tractus olfactorius (T.Of.) and enters the deep plexus (D.P.), establishing contacts with cells a, b, and c. These fibers might represent either axons of cells placed rostrally and coursing rostrocaudally through the deep plexus (axons of horizontal cells reproduced in Fig. 29, as well as other cells of the fourth layer) or efferent olfactory bulb fibers. This latter assumption is based on observa-

tions made in experimental material showing that after interruption of the tractus olfactorius some degenerating fibers could be traced to the deep layers of the cortex piriformis (see Cat No. 8 and Fig. 12 of the experimental series of the present work). A third olfactory inflow passes directly to the tuberculum olfactorium (Tu.Olf.), whose cells contribute axons to the medial forebrain bundle.

Cells a, b, and c of Fig. 43D could convey olfactory impulses to the medial forebrain bundle. Cell a corresponds to the type described as a variety of polygonal cell with dendrites radiating in all directions (see also Fig. 40, g, l, and Fig. 43B). The axon of this cell courses through the deep plexus, entering later into the medial forebrain bundle (M.F.B.). The axon of cell b of Fig. 43D corresponds to the red fiber labeled 9 and its collaterals, reproduced in Fig. 43B; this cell possesses an ascending collateral to the fibrillar layer, another collateral which

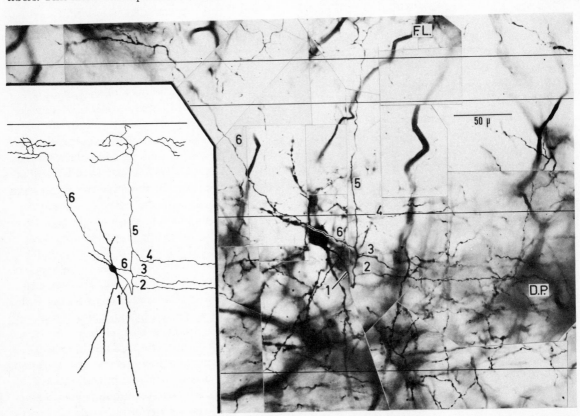

Fig. 41. Mouse 41L, 4 days old. Mosaic photomicrographic reconstruction of a stellate cell with ascending axon in layer IV (D.P.) of the cortex piriformis. A diagram of the cell is shown at the left. Golgi method. (From Valverde, 1963a.)

establishes contact with cell *c*, and a main axonal branch which courses first through the deep plexus (D.P. in Fig. 43*D*) and later in the tuberculo-piriform system, where it sends a collateral to the tuberculum olfactorium (Tu.Olf.), and finally enters into the medial forebrain bundle (M.F.B.). Cell *c* of Fig. 43*D* (see also Fig. 38, *g*, and Fig. 41) possesses an ascending axon reinforcing the fibrillar layer and collaterals which distribute through the deep plexus. Probably this type of cell corresponds to the last variety of short axis cylinder cells described by O'Leary (1937) in his layer III.

From this schema one gains the impression that multiple and different circuits can be traced through the different types of cells described previously in the cortex piriformis. As one example,

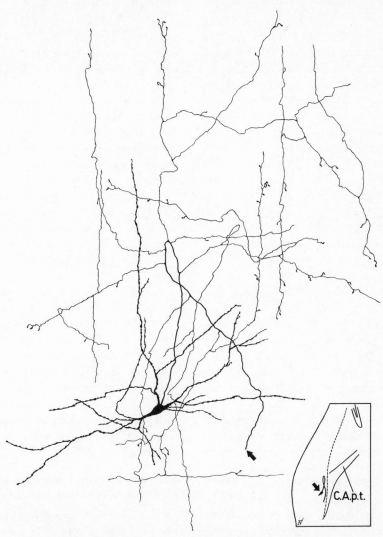

Fig. 42. Rat 43R, albino, 9 days old. Horizontal section showing one of the horizontal cells of the cortex piriformis in front of the commissura anterior, pars temporalis (C.A.p.t., see inset diagram). This type of cell is particularly abundant in the depth of the cortex (see Fig. 29); on account of the disposition of its axon and collaterals running in a rostro-caudal direction, it is assumed that this type of cell relates distant parts of the cortex piriformis. Golgi method.

impulses coursing through the blue fiber of the fibrillar layer (F.L. in Fig. 43D) might pass through the apical dendritic branch of cell c to the body which in turn activates, by its ascending axon, the apical dendritic branch of cell b; from the body of the latter, impulses could be conducted to the fibrillar layer by its ascending collateral, might synapse again with cell c, or might course through the deep plexus (D.P.) to the medial forebrain bundle (M.F.B.). It is assumed that impulses traveling through the tractus olfactorius reach the two plexuses of this cortex. Cells like those labeled c in Fig. 43D (see also Fig. 38, g) convey the olfactory impulses caudally (toward the entorhinal cortex), and, assuming that there exists a rostrocaudal row of these cells, a zigzag (fibrillar layer–deep plexus–fibrillar layer) conduction may take place. This form of possible conduction is reinforced by impulses traveling horizontally through the deep plexus; the susbtratum of this conduction is represented by the horizontal cells described in the deep plexus (see Figs. 29 and 42) as well as by horizontal collaterals, furnished in this layer, by the axons of a great number of cells of the cortex piriformis. Electrophysiologic evidence of a multisynaptic pathway throughout the prepiriform cortex to the hippocampus has been described by Cragg (1960, 1961b).

In the mouse, O'Leary (1937) was able to stain several types of short-axis cylinder cells of the cortex piriformis that I have not observed in the present study. Conversely, in my Golgi preparations, I could follow other axonal patterns not observed by O'Leary (1937). Needless to say, the Golgi method shows different stained structures from specimen to specimen and consequently from author to author. Only the combination of partial observations will show a clear and complete picture of a given structure. O'Leary (1937) made clear that the deep pyramids would be responsive to olfactory stimuli through intercalated superficial pyramids. He also stated that the deep polymorphic cells with dendrites restricted to the depth of the cortex could be activated only through the superficial or deep pyramids.

According to O'Leary (1937), short-axis cylinder cells would play a major role in reverberating circuits and in synchronizing the activity of groups of pyramidal cells.

In addition to that, I would emphasize that direct influences of the olfactory input upon deep cells of the cortex piriformis take place, provided that many cells of the deep piriform layers possess dendrites reaching the fibrillar layer and that the many ascending axons observed in the present investigation belonging to deep cells would be able to act upon other ascending dendrites, of other deep cells, reaching also the fibrillar layer. Generalizing, it could be assumed that the short-axis cylinder cells described by O'Leary (1937) form a series of restricted reverberating circuits of probably slow propagation and that the zigzag type of conduction described above would have a more widespread and faster propagation, reinforced by horizontally running axons, upon distant parts of the cortex.

To summarize, the cortex piriformis (area piriformis anterior and medialis) acts upon the six following organizations:

(1) The contralateral olfactory bulb, nucleus olfactorius anterior, and rostral piriform cortex through the commissura anterior, pars bulbaris (see pp. 54–59);

(2) The medial forebrain bundle (see pp. 83–86);

(3) The regio praeoptica through the area amygdaloidea anterior (see pp. 63 and 83);

(4) The contralateral area piriformis medialis through the commissura anterior (see pp. 59–61);

(5) Caudally throughout the intracortical zigzag and horizontal chains just described (see above) to the area entorhinalis;

(6) The amygdaloid complex.

The first five of these possibilities have been discussed; the last one will be treated in the next paragraph.

THE PIRIFORM-AMYGDALOID RELATIONS.
OLFACTORY CONNECTIONS OF THE AMYGDALA.

The cortex piriformis has been claimed to be one of the main sources of amygdaloid afferents as shown by several workers (Johnston, 1923; Sprenkel, 1926; Hilpert, 1928; Gurdjian, 1928; Kappers, Huber, and Crosby, 1936; Mittelstrass, 1937; Fox, 1940; Pribram, Lennox, and Dunsmore, 1950; Macchi, 1951; Lammers and Lohman, 1957).

The amygdaloid complex is covered ventro-

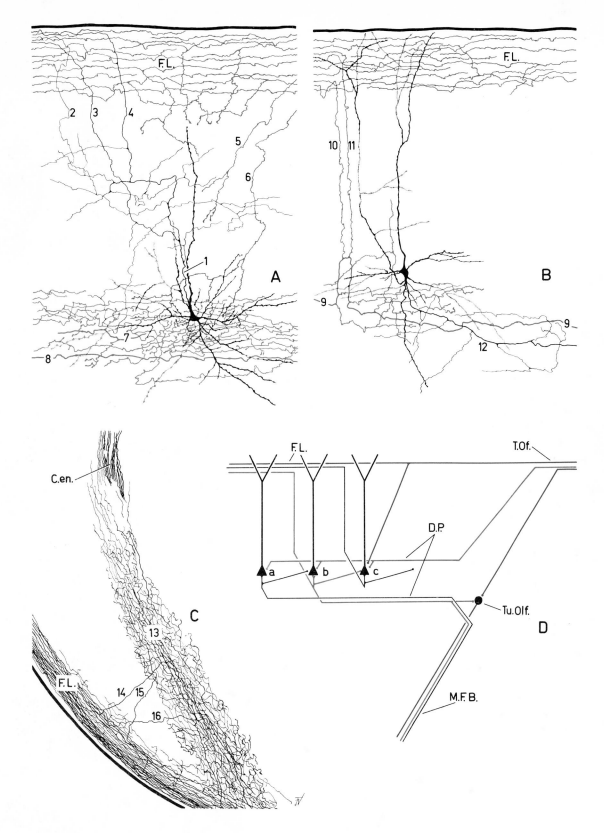

Fig. 43. (*A*) Mouse 17, 3 days old. A cell of the deep plexus of the cortex piriformis with ascending axon and branches contributing to the fibrillar layer. Golgi method.

(*B*) Mouse 19, 3 days old. A cell of the deep plexus of the cortex piriformis. The axon of this cell (*12*) and the axon labeled *9* (belonging to another cell of the cortex piriformis) could be traced to the tuberculum olfactorium. Golgi method.

(*C*) Rat 41R, albino, 4 days old. Transverse section through the cortex piriformis, showing a zone of confluence of the deep plexus (*13*) and the fibrillar layer (F.L.). Golgi method.

(*D*) Diagrammatic representation of the conduction of impulses through the cortex piriformis.

laterally by a considerable extension of the cortex piriformis (the periamygdaloid cortex). In frontal sections the limits between the two structures (Fig. 45) are marked by the ventromedial portion of the capsula externa (C.en.) lying between the cortex and the nucleus amygdaloideus lateralis (A.l.). More medially, no sharp boundary exists between the cortex and the nucleus amygdaloideus basalis (A.b.), and still more medially the nucleus amygdaloideus corticalis (A.co.) fuses imperceptibly with the ventral parts of the nucleus amygdaloideus basalis.

The part of the amygdala covered directly by the cortex piriformis comprises, thus, the ventro-lateral surface of the nucleus amygdaloideus lateralis and the ventral limits of the nucleus amygdaloideus basalis. Through these regions the cortico-amygdaloid interconnections take place. With some minor variants, they appear to be similar in the rat and the cat.

In the cat (Fig. 45) a relatively thick sheet of fibers lies between the nuclei amygdaloideus lateralis (A.l.) and basalis (A.b.). This plate of fibers conveys a great number of axons of cells of the underlying cortex piriformis. As shown by this figure, the axons of cells d, e, and f pierce the deep plexus (D.P.) perpendicularly and emit several collaterals, either medially to the nucleus amygdaloideus basalis (collateral 2d of the axon 1d and collateral 2e of 1e) or laterally to the nucleus amygdaloideus lateralis (collateral 2 of fiber 1).

Farther laterally the capsula externa (C.en.) interposes between the axons of cells of the cortex piriformis, compelling these axons to turn laterally and course through the capsula until, condensed in several little fascicles, they enter the nucleus amygdaloideus lateralis (A.l.). This can be clearly seen in the course followed by the axon of cell c of Fig. 45.

More laterally the axons of cells a and b entered the capsula externa where they could be followed dorsally. The little fascicles piercing the nucleus amygdaloideus lateralis become more and more displaced laterally until they fuse with the capsula externa. On the basis of these observations, it could be assumed that these fascicles represent aberrant fibers of the capsula externa that shorten their pathway to the sublenticular region by coursing through the nucleus amygdaloideus lateralis.

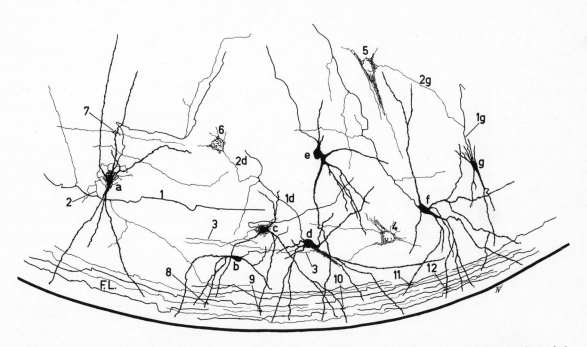

Fig. 44. Cat 30, embryo. Transverse section through the fibrillar layer (F.L.) and overlying portion of the periamygdaloid cortex at the level of transition with the nucleus amygdaloideus corticalis. Golgi method.

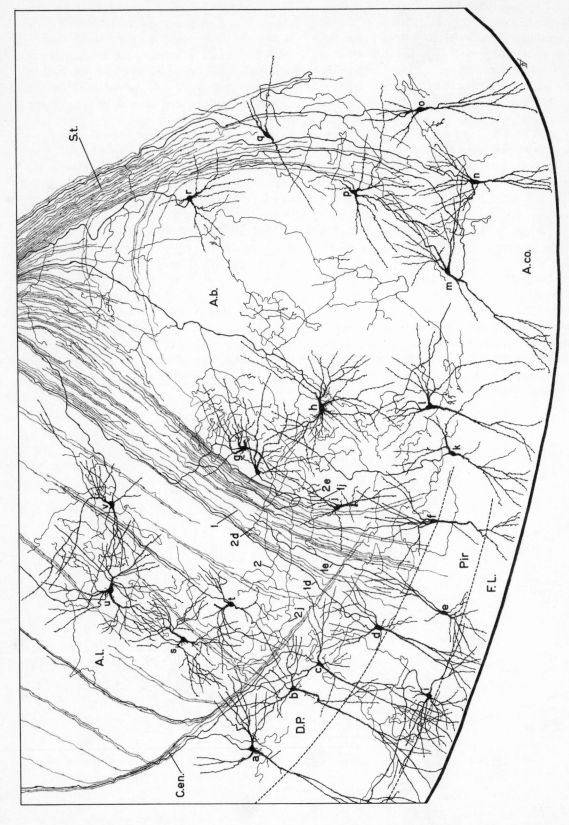

Fig. 45. Cat 31, 1 day old. Transverse section through the amygdaloid complex, showing the organization of the piriform-amygdaloid relations. Golgi method.

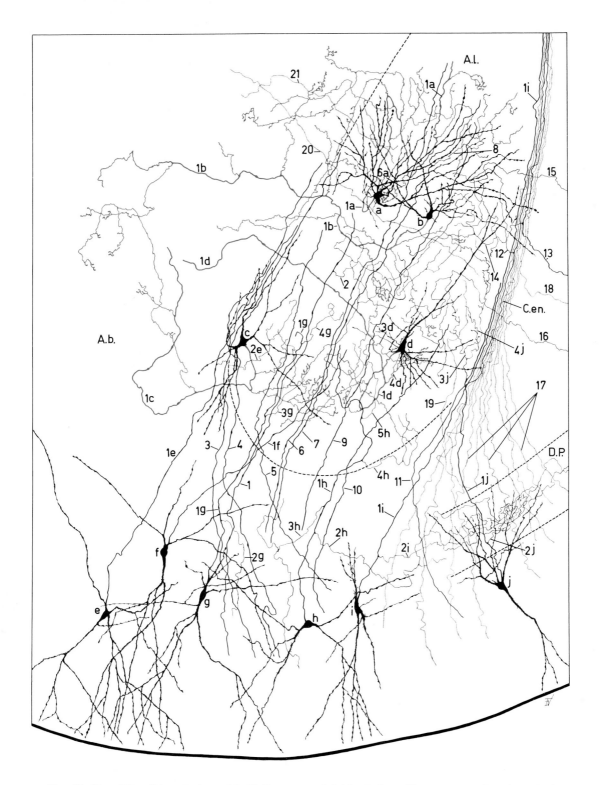

Fig. 46. Rat 43L, albino, 9 days old. Piriform-amygdaloid relations. Transverse section through the nucleus amygdaloideus lateralis. Axons of pyramidal cells of the underlying periamygdaloid cortex enter the nucleus by its ventral aspect. Axons of amygdaloid cells (drawn in red) project medially. Golgi method.

Many axons of cells of the nuclei amygdaloideus lateralis and basalis course either with the little fascicles piercing the former or among the plate of fibers separating the two nuclei, as illustrated by the course outlined of the axons of cells s, t, u, and v (Fig. 45) of the nucleus amygdaloideus lateralis (A.l.) and for the axon of cell g of the basalis (A.b.).

The piriform-amygdaloid connections in the rat are similar. Figure 46 represents a transverse section showing the connections existing between the cortex piriformis underlying the nucleus amygdaloideus lateralis and this nucleus (A.l.). Medially, the axon 1e of cell e ascends directly to the medial boundary of the nucleus amygdaloideus lateralis; the axon 1e joins fibers 3 and 4 belonging to other cortical cells not represented; they follow a long course through the zone separating the nuclei amygdaloideus lateralis (A.l.) and basalis (A.b.).

The axons 1f, 1g, and 1h (Fig. 46), belonging respectively to cells f, g, and h, enter through the ventral surface of the nucleus lateralis. Other axons labeled 1, 5, 6, 7, 9, and 10 (belonging to piriform cells not represented) behave in the same manner; they enter the nucleus amygdaloideus lateralis grouped in small fascicles, and ascend dorsolaterally through this nucleus. More laterally the axons of piriform cells enter the capsula externa (C.en.) directly, as shown by axons 1i and 1j of cells i and j respectively. The same occurs with the axon labeled 11 and the group of fibers labeled 17.

In the rat these three systems of fibers connecting the cortex with the nucleus amygdaloideus lateralis (Fig. 46) might be called medial (axons 1e, 3 and 4), intermediate (axons 1f, 1g, 1h, 1, 5, 6, 7, 9, and 10), and lateral (axons 1i, 11, and group 17) piriform-amygdaloid systems. These three groups of fibers give off numerous collaterals to the nucleus amygdaloideus lateralis. Several of these collaterals are illustrated in Fig. 46:

(1) From the medial group, collateral 2e of axon 1e;

(2) From the intermediate group, collaterals 3g and 4g of axon 1g; 4h and 5h of axon 1h; 2 of axon 1; 8 of axon 7.

(3) From the lateral group, collaterals 3j and

4j of axon 1j; the fiber labeled 12 which is a collateral of axon 11. Other fibers of this lateral group approach the capsula externa along its lateral border: fiber 13 enters the capsula externa giving off collateral 14 to the nucleus amygdaloideus lateralis; fibers 15 and 16 pierce the capsula entering directly the nucleus amygdaloideus lateralis, where they give off several collaterals; while fiber 18 and the group of fibers 17 do not appear to emit collaterals to this nucleus; however, it should be observed that dendrites of cells a, b, and d penetrate into the capsula externa, where synaptic contacts probably are made.

The piriform-amygdaloid connections also are reproduced in Fig. 47, which is a frontal section through the amygdaloid complex of the adult cat stained by the Heidenhain method. The system 1 of Fig. 47 represents the medial group of the piriform-amygdaloid fibers traced previously in Figs. 45 and 46. This group of fibers ascends dorsally, turns laterally surrounding the medial limits of the nucleus amygdaloideus lateralis (A.l.), and forms a flattened system of fibers that I have called the capsula intermedia (C.im.); later it fuses with the capsula externa (C.en.) along the dorsal edge of the nucleus amygdaloideus lateralis. As mentioned previously, this system of fibers conveys axons of piriform cells as well as axons of cells of the amygdaloid complex connecting the cortex piriformis and ventral amygdala with the more dorsal parts of the latter.

Reference is made to the description of Cat 4 of the experimental series, in which the striking resemblance of the degeneration traced in that experiment to some of the systems of fibers drawn in Fig. 47 is obvious. Comparing Fig. 47 with Fig. 6, the system labeled 1 in the former (also labeled capsula intermedia, C.im.) represents the fantail system traced in sections 2 and 3 of Fig. 6.

The intermediate (group 2 of Fig. 47) and lateral (included in the capsula externa, C.en., Fig. 47) piriform-amygdaloid group of fibers also were damaged by the lesion in Cat 4.

In summary, the following conclusions have been drawn from these observations:

(1) The periamygdaloid cortex is related with the amygdala by three systems. The medial (the largest) group of fibers ascends dorsally through

the basal amygdaloid nucleus and also between this and the lateral amygdaloid nucleus. It has been called the capsula intermedia or the fantail system. The intermediate or diffuse system of fibers is composed of little fascicles that pierce the nucleus amygdaloideus lateralis, and the lateral system fuses with the latter and ascends through the capsula externa, from which some fibers leave to enter the nucleus amygdaloideus lateralis.

(2) These three systems convey piriform-amygdaloid fibers as well as fibers relating the ventral parts of the amygdala with its dorsal portions. As observed in Cat 4, no degeneration could be traced beyond the amygdala.

(3) The medial piriform-amygdaloid system extends rostrally and fuses imperceptibly with the rostral projection system of the amygdala as observed in Cat 4, section 1 (Fig. 6).

Although several axons have been traced from the cortex piriformis directly to the nucleus amygdaloideus basalis, in agreement with Fox (1940) my Golgi sections have revealed that the nucleus amygdaloideus lateralis is the main receptor site for piriform axons. As shown in Fig. 46, the cells of the latter possess axons directed medially (axon *1a* of cell *a* and axons *1b*, *1c*, and *1d* of cells *b*, *c*, and *d* respectively). We can conclude, therefore, that this nucleus acts in part as a link between the cortex piriformis and the medial parts of the amygdala where the stria terminalis originates. Comparing Fig. 46 with Fig. 49, it can be observed that the nucleus amygdaloideus basalis receives olfactory impulses through the intermediate links represented by cells of the nucleus amygdaloideus corticalis (Fig. 49, A.co., axons of cells *v* and *x*) and that the nucleus amygdaloideus

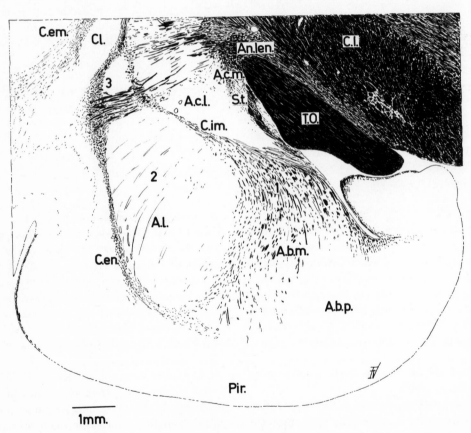

Fig. 47. Transverse section through the amygdaloid complex of an adult cat, showing the fibrillar structure of the region. Heidenhain stain.

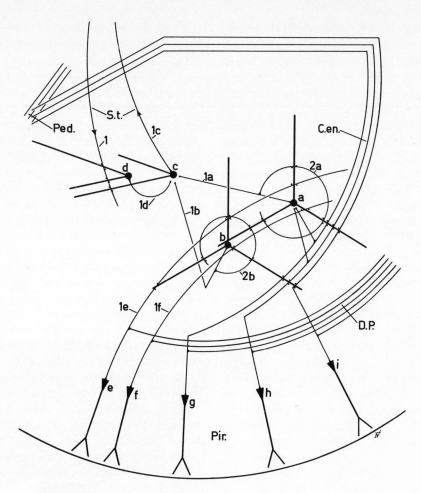

Fig. 48. Diagrammatic representation of the piriform–amygdaloid–stria terminalis circuit.

lateralis receives olfactory projections through the periamygdaloid cortex (cells *e*, *f*, *g*, *h* and *j* of Fig. 46).

These piriform-amygdaloid connections form a link in a complex circuit relating the olfactory fibers with the diencephalon through the amygdala and stria terminalis. This has been schematized in Fig. 48. In this figure *e*, *f*, *g*, *h*, and *i* represent cells of the periamygdaloid cortex; *a* and *b* are two cells of the lateral portions of the amygdala; *c* and *d* represent two cells of the medial parts of the amygdala.

In the schema of Fig. 48 cells *a* and *b* have been drawn as each possessing three radiating dendrites and one axon (*1a* and *1b*) directed medially and emitting a collateral (*2a* and *2b*). These

cells represent a schema of the cells *a*, *b*, *c*, and *d* drawn in the nucleus amygdaloideus lateralis in Fig. 46; the dendrites of *a* and *b* (Fig. 48) that penetrate the capsula externa (C.en.) illustrate the protrusion of some dendrites of cells *a*, *b*, and *d* into the capsula externa (Fig. 46, C.en.); the collaterals *2a* and *2b* (Fig. 48) also illustrate the complex collateral apparatus of cells *a* and *b* of Fig. 46 which ramifies within the limits of the dendritic field of these cells.

In Fig. 48 the axons of cells *e* and *f* (*1e* and *1f*) enter the amygdala and approach cells *a* and *b*, establishing synaptic contacts with their dendrites (represented by dots where the drawing lines intersect). The axons *1e* and *1f* each give off a collateral that enters the deep plexus (D.P.) of the cortex piriformis; this represents a common feature observed in some axons of cells of the cortex piriformis as, for example, collaterals *2h* and *2i* in Fig. 46. The axons of cells *g*, *h*, and *i* (Fig. 48) each give off a collateral to the deep plexus. These axons enter the capsula externa (C.en.), where they establish synaptic contact with the dendrites of cells *a* and *b*, penetrating into the capsula.

Therefore, cell *c* (Fig. 48) could be driven, through axons *1a* and *1b*, by the activity of cells of the cortex *e* to *i*; cell *c* contributes to the stria terminalis (S.t.) via its axon *1c*; in turn, cell *c* also could be activated through the axon *1d* of cell *d*, by the amygdalopetal part of the stria terminalis (fiber *1*).

In my Golgi material I could not determine where the axons of piriform cells terminate. Presumably, they enter the sublenticular region either to be incorporated in the projection system of the temporal lobe or to pass caudally to the hippocampus. The degeneration traced after the lesion of Cat 4 develops completely within the amygdaloid region, but in Cat 5 some fibers, belonging to axons originating in the cortex piriformis, could be traced to the ventral hippocampus through the lateral parts of the amygdala. Presumably, part of the piriform-amygdaloid relations described for Fig. 46 would bypass caudally the nucleus amygdaloideus lateralis to enter the hippocampus.

The present studies suggest the existence of an "olfactory bulb-stria terminalis" pathway as follows: fibers of the tractus olfactorius lateralis project to the nucleus amygdaloideus corticalis; the latter projects to the basalis, which, in turn, sends its fibers either to the lateralis, where some fibers of the stria terminalis originate (Fig. 52), or to the centralis and medialis, where the stria also originates. Further observations suggest that the following represents another "olfactory bulb–stria terminalis" circuit: fibers of the tractus olfactorius lateralis project to the periamygdaloid cortex; the latter projects to the lateralis, whose cells send their axons to the stria terminalis either directly (see Fig. 52, axon of cell *a*) or through relays in the centralis and medialis. However, it should be added that there exist amygdalofugal pathways besides the stria terminalis, which project either

rostrally through the rostral projection system of the amygdala or caudally, toward the hippocampus.

The above observations allow us to consider that, functionally, both the nucleus amygdaloideus corticalis and the adjoining periamygdaloid cortex act similarly, subserving olfacto-amygdaloid connections; consequently, the former should be referred to as a cortical formation rather than a nucleus of the amygdaloid complex. A similar point of view should be held for the nucleus tractus olfactorii lateralis, which is usually considered an amygdaloid nucleus on account of its well established stria terminalis contributions; its position and structure also indicate that it might be considered a cortical formation like the nucleus amygdaloideus corticalis and the periamygdaloid cortex. The nucleus tractus olfactorii lateralis and nucleus amygdaloideus corticalis should be included in the receptive olfactory cortical area together with the periamygdaloid cortex and other formations, rather than considered as subdivisions of the amygdala. Therefore, if one excludes the nucleus amygdaloideus corticalis and the nucleus tractus olfactorii lateralis from the amygdaloid complex, the amygdala takes on a different functional significance and obtains the position whereby it is able to have the same relation with the olfactory inflow as with other extraolfactory cortical or subcortical influences.

The olfactory connections of the amygdala have been extensively studied. Rose and Woolsey (1943), Fox, McKinley, and Magoun (1944), and Kaada (1951) recorded evoked potentials to electric stimulation of the olfactory bulb in the nucleus amygdaloideus corticalis; such responses are mediated by the collaterals of the tractus olfactorius lateralis to the cells of the cortical nucleus, in agreement with several workers (Johnston, 1923; Gurdjian, 1928; Loo, 1931; Humphrey, 1936; Jeserich, 1945; Lauer, 1945; Clark and Meyer, 1947; Meyer and Allison, 1949; Adey, 1953; Ward, 1953; Lohman and Lammers, 1963; Lohman, 1963). Although it is generally assumed that direct impulses from the olfactory bulb reach the cortico-medial complex, it appears that the bulk of these connections are established mainly with the nucleus amygdaloideus corticalis and

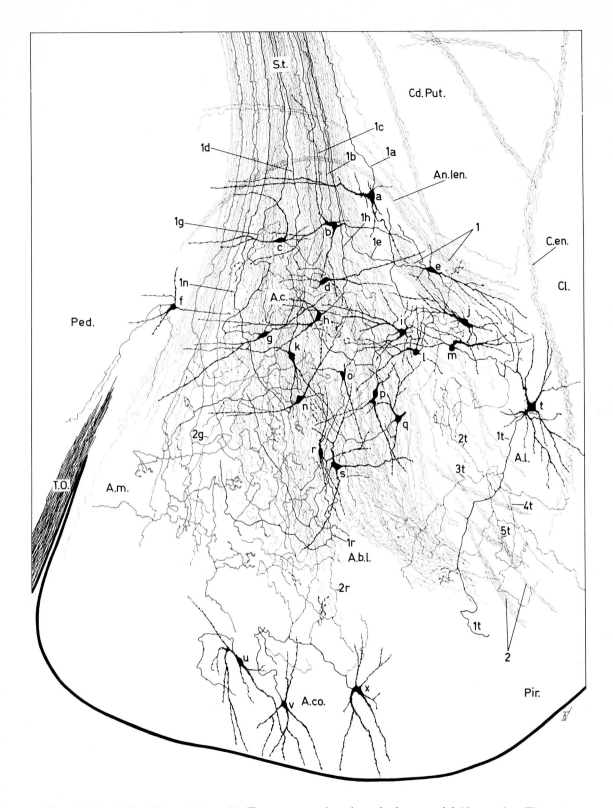

Fig. 49. Rat 41L, albino, 4 days old. Transverse section through the amygdaloid complex. The stria terminalis (S.t.) forms the background of the drawing. Several cells send their axons to the stria. Amygdalopetal stria fibers have been drawn in red. Note the horizontal disposition of some dendrites. Golgi method.

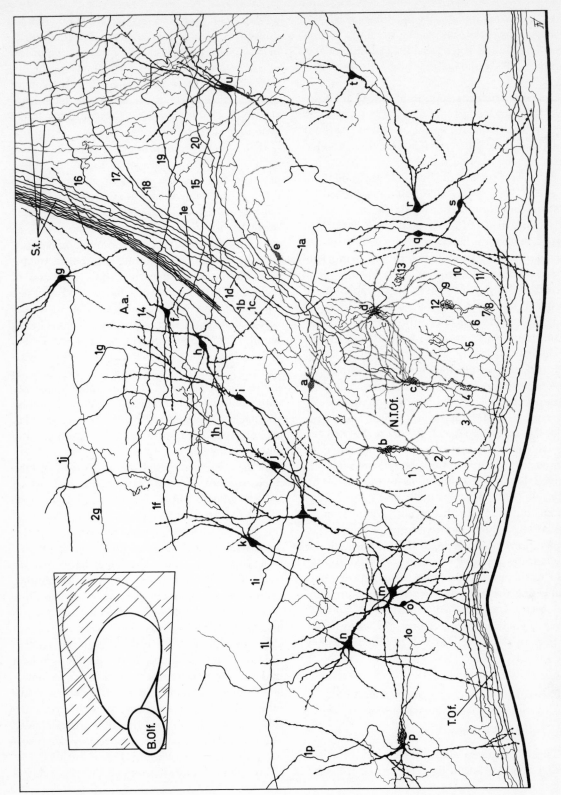

Fig. 50. Rat 45L, albino, 9 days old. Sagittal–oblique section (as indicated in the inset diagram) through the nucleus tractus olfactorii lateralis (N.T.Of.). Some fibers of the stria terminalis (S.t.) originate in cells of this nucleus. Golgi method.

that few, if any, direct olfactory fibers reach the nucleus amygdaloideus medialis, that is, the amygdala proper, as Jeserich (1945), Lohman and Lammers (1963), and Lohman (1963) have pointed out.

In Cat 8, after interruption of part of the tractus olfactorius lateralis, no fibers were seen entering the amygdala. According to Kölliker (1896) and Cajal (1911), the amygdala does not receive direct olfactory connections.

Berry, Hagamen, and Hinsey (1952) and Hugelin *et al.* (1952) showed that potentials evoked by olfactory-bulb stimulation could be obtained in the whole extent of the amygdala; this lends support to the assumption that the remainder of the amygdala is activated either through the nucleus amygdaloideus corticalis or through the periamygdaloid cortex. Moreover, these responses are similar in character to those recorded after stimulation of other cortical or subcortical structures, as substantiated by several physiologic studies to be discussed below.

Little information has been gained from the present observations concerning amygdalo-piriform connections. In Fig. 49 the collateral *2r* of the axon *1r* of cell *r* approaches the nucleus amygdaloideus corticalis (A.co.); the axon *1t* of cell *t* of the same figure appears to enter the periamygdaloid cortex, but it could not be followed farther. Amygdalo-piriform connections were described by Johnston (1923), Hilpert (1928), and Lauer (1945) among others. From the periamygdaloid cortex caudalward, short multisynaptic pathways described previously (see pp. 85–86), confirming the observations reported by Cragg (1961a, b), extend to the area entorhinalis. Evidence of fibers connecting the amygdala with the periamygdaloid cortex was observed in Cats 1, 5, and 24. Their significance concerning the amygdalo-hippocampal connections will be discussed later (see pp. 110–112).

THE AMYGDALOID COMPLEX, STRIA TERMINALIS, AND FASCICULUS LONGITUDINALIS ASSOCIATIONIS

The Amygdaloid Complex

The amygdala is usually divided into a cortico-medial complex and a basolateral one which in phylogenetic evolution have retained an essential uniformity. The former, composed of the central, medial, and cortical nuclei, represents the primitive lateral area of fishes in which part of the lateral olfactory tract terminates. The basolateral complex, composed of the basal and lateral nuclei, represents the most recent acquisition in the evolution of the amygdala.

The olfactory relations of the amygdala have been a matter of controversy for a long time; however, this question has been solved partly on the basis of comparative anatomical studies. It appears that Kölliker (1896) and Cajal (1911) were the first to deny its olfactory relations. It was observed in cetaceans (anosmatic mammals) that the amygdala (excluding the nucleus tractus olfactorii lateralis) does not show any sign of regression (Breathnach and Goldby, 1954). In these aquatic mammals the great size attained by the amygdala is paralleled by the high degree of temporalization (greater than that reached in the human brain). This speaks in favor of the existence of a close developmental relation between the temporal lobe and the amygdala, especially with the basolateral complex as substantiated by Whitlock and Nauta (1956). In cerebral arhinencephalic malformations the amygdala is normally developed (Yakovlev, 1959).

If the nucleus amygdaloideus corticalis and the nucleus tractus olfactorii lateralis are included in the amygdaloid complex, the amygdala must be considered an olfactory center of the first order, but if these structures are considered as special formations of the periamygdaloid and prepiriform cortices (as their histologic structure indicates), the amygdala, excluding such formations, must be considered a subcortical center integrating various sensory modalities of which olfaction is one. If the foregoing account is accepted, the nucleus amygdaloideus corticalis and the nucleus tractus olfactorii lateralis should be grouped under a common term: the *direct olfactory projection area* (rhinencephalon proper) with the nucleus olfactorius anterior, the cortex praepiriformis, the tuberculum olfactorium, the area amygdaloidea anterior, and that part of the periamydaloid cortex which receives olfactory fibers directly.

Abundant literature exists on the cytoarchitecture of the amygdaloid complex in various mammals (Kölliker, 1896; Völsch, 1906, 1910; Johnston, 1923; Sprenkel, 1926; Hilpert, 1928; Smith, 1930; Loo, 1931; Humphrey, 1936; Kappers, Huber, and Crosby, 1936; Young, 1936; Mittelstrass, 1937; Brockhaus, 1938; Fox, 1940; Crosby and Humphrey, 1941, 1944; Jeserich, 1945; Lauer, 1945; Brodal, 1947; Karibe, 1961; Valverde, 1962; Koikegami, 1963a, b).[7] These studies have traced the boundaries and described the cytoarchitectonics of the different nuclei, as well as discussing some amygdaloid connections. The following description, based on Golgi observations, presents a general picture of the cells of the amygdala.

Frontal Golgi sections of the mouse show a clear demarcation between the basolateral complex and the medial group; in the latter the cells are loosely grouped and no clear delimitation between central and medial nuclei can be established; the cells of both the central and the medial are medium sized. The cells of the basolateral complex have a profuse dendritic apparatus; those of the basalis are medium sized; those of the lateralis in the mouse are the largest in the amygdaloid complex. This is in marked contrast to previous observations made in the cat in which the giant cells of the amygdala are confined to the magnocellular portion of the nucleus basalis.

Several examples of cells of the amygdala in the cat are illustrated. In Fig. 45, s, t, u, and v are examples of cells of the nucleus amygdaloideus lateralis; their bodies are irregular and have profuse dendritic processes. Cells g, h, and j of the basalis have a similar morphology whereas the cells p, q, and r located medially in the amygdala have scanty dendrites and small-sized bodies.

In the rat, as in the mouse, the largest of the amygdaloid cells are located in the nucleus amygdaloideus lateralis (Fig. 46, a, b, c, and d; Fig. 49, t; Fig. 52, a). Their dendritic ramifications appear to radiate in all directions; however, the dendrites of cells a and b (Fig. 46) tend to group within the area of the abundant collaterals which are furnished by the axon 1a throughout its N-like course. Frequently, as described previously, dendrites of cells of the lateralis penetrate the capsula externa, where they make synaptic contacts (see dendrites of cells a, b, and d of Fig. 46, of cell t of

Fig. 49, of the cell of Fig. 51A, and of cell a of Fig. 52).

It is difficult, on the basis of Golgi material, to outline the boundaries of the different amygdaloid nuclei. In Fig. 49, cells o, p, q, r, and s belong to the nucleus amygdaloideus basalis; some of them have dendrites radiating in all directions; fusiform cells with two opposite bushes of dendrites are frequently seen (cells p and r). Figure 49 shows an outstanding difference between cells of the basolateral group and those of the nucleus amygdaloideus centralis (A.c.); in the latter the dendrites are oriented perpendicular to the fibers of the stria terminalis, S.t. It seems that these dendrites receive either afferent or efferent amygdaloid impulses traveling through fibers of the stria terminalis.

Concerning the axonal apparatus, the amygdaloid cells can be classified into two main groups: long projecting ones and intermediate axoned cells.

(1) *Long projecting cells*. This type of cell includes those whose axons enter the stria terminalis. Before entering the stria, the axon of this type of cell may emit several collaterals (cells h, p, q, and r of Fig. 45; axons 1d, 1g, 1h, and 1n of cells d, g, h, and n of Fig. 49) or may enter the stria terminalis directly without giving off any collaterals (axons 1a, 1b, 1c, and 1e of cells a, b, c, and e of Fig. 49; the cell of Fig. 51B).

(2) *Intermediate-axoned cells*. This type includes the remainder of the amygdaloid cells whose axons do not contribute to the stria terminalis. These cells have axons of intermediate length and numerous collaterals (Fig. 51A).

Excluding a scanty number of cells with axons without collaterals, I gained the impression that the rest of the axons of amygdaloid cells give off many collaterals, which arborize profusely among the surrounding cells and contribute to the formation of the amygdaloid plexus. This profuse network of collateral fibers on one side interconnects the different amygdaloid nuclei and on the other side extends forward to the area amygda-

[7]There exists an abundant literature on the amygdaloid complex written in Japanese. A review of these reports and new observations has recently been published in English by Koikegami (1963b).

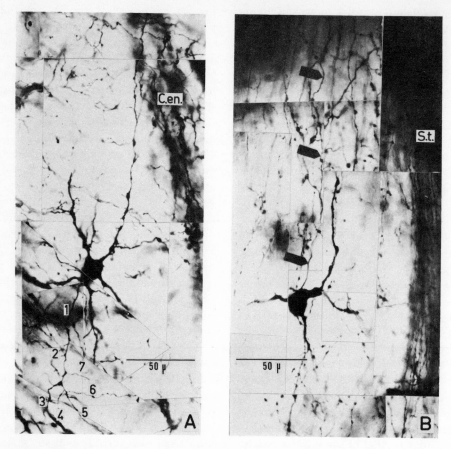

Fig. 51. Rat 41L, albino, 4 days old. Mosaic photomicrographic reconstruction of the two kinds of amygdaloid cells.

(A) A cell in the nucleus amygdaloideus lateralis near the capsula externa (C.en.) with axon (1) branching into several collaterals (2 to 7).

(B) A cell with axon (indicated by three arrows) entering the stria terminalis (S.t.) directly, without emitting collaterals. Golgi method.

loidea anterior and regio praeoptica, forming a rather diffuse multisynaptic chain system which can be considered as the amygdalofugal part of the *longitudinal association bundle* of Johnston (1923) or the *rostral projection system of the amygdala*. As mentioned previously, in the mouse and rat, this system is composed of short neuron chains.

The main axonal branch of the intermediate type of cell extends moderately far from the cell body, ending as a fine fiber and emitting collaterals all along its course. The distance covered by these axons is difficult to measure; nearly complete axonal apparatuses were observed in Fig. 46, cells *a* and *d*, and in Fig. 49, cell *t*; in these

cells it was impossible to follow their axons to their terminations. It is difficult to decide which cells do not send their axons to the stria terminalis. Axons were frequently observed following tortuous courses and emitting several collaterals before entering that bundle.

Cell configuration does not allow classification as to whether a given amygdaloid cell belongs to the stria-projecting type or to the intermediate-axoned type; thus, it is necessary to verify *de visu* the entrance of their axons into the stria in order to classify them as belonging to the first type described. For this reason, it became impossible to decide which of the cells reproduced in Figs. 45

and 46 belong to the intermediate-axoned type, on account of the possibility that their axons might finally enter the stria. In Fig. 49 it is reasonable to consider that cells *f, k, o,* and *t* do not send their axons to the stria and, therefore, they belong to the intermediate-axoned type. Cells *i, j, l, m, p, q, r,* and *s* cannot be classified in any sense.

The second type of cell just described fits neither into the Golgi type II nor into the Golgi type I cells; I prefer to label them as cells of *intermediate-axoned type* on account of the distance covered by their axons and collaterals, which, on the one hand, is never as long as that reached by the long projecting cells but, on the other hand, exceeds the field covered by the axon of a Golgi type II cell (authentic Golgi type II cells are those reproduced in Fig. 31 in the regio praeoptica, R.P.O., and nucleus accumbens septi, Acb.); however, it should be borne in mind that this is a personal point of view.

The resemblance of these intermediate axoned cells to those bearing the same name, which I described previously in the lateral reticular formation of the brain stem (Valverde, 1961), appears to lend support to this classification.

The stria terminalis

Although the pattern of origin of the stria terminalis within the amygdaloid complex is not well established, there is general agreement that in the cat this bundle originates mainly in the corticomedial complex (Fox, 1940, 1943). The basolateral complex does not send fibers to the stria (Gloor, 1955a). It is generally assumed that this latter complex, by short intra-amygdaloid connections, feeds the corticomedial one and thence the stria terminalis.

In the mouse the pattern of origin of the stria terminalis differs somewhat with respect to that observed in the cat; in the former I observed that the central and medial amygdaloid nuclei and also cells of the basal and lateral nuclei contribute axons to the stria. Figure 52 is a transverse section through the amygdaloid complex in a 20-day-old mouse. A thin amygdalopetal fiber of the stria terminalis (S.t.) divides into two branches which, in turn, emit other collaterals branching

off in the mediobasal region of the amygdala. One of the final branches of the thin amygdalopetal fiber of the stria terminalis emits short terminals to the cell labeled *d*; the axon of the latter enters the nucleus amygdaloideus lateralis (A.l.). A large cell in the latter nucleus (labeled *a*) sends the axon to the stria terminalis (S.t.). A circuit consisting of *amygdalopetal stria fibers–medial amygdaloid regions–lateral amygdaloid regions–amygdalofugal stria fibers* (see inset diagram of Fig. 52) was proposed previously (Valverde, 1962), based on this and similar observations.

In Fig. 45 it can be seen that cells *m, n,* and *o,* belonging to the nucleus amygdaloideus corticalis (A.co.), send their axons to the stria terminalis. On the basis of my Golgi sections, I was able to observe in two main nuclei cells contributing to the efferent portion of the stria terminalis: the nucleus amygdaloideus centralis (Fig. 49, cells *a, b, c, d, e, g, h,* and *n*) and the nucleus tractus olfactorii lateralis (Fig. 50, N.T.Of., cells *a, b, c, d,* and *e*).

The possibility cannot be discarded that the periamygdaloid and temporal cortices are sources of efferent stria fibers. With regard to this assumption, in Fig. 49 labels *1* and *2* indicate several small bundles coming from the temporal cortex and approaching the stria terminalis; these fibers would form in part the ansa lenticularis (An.len.), but it may be possible that some of them enter the stria terminalis.

Figure 50 has been drawn from a section taken from a rat's brain cut at an angle of about 45 degrees with respect to the frontal or sagittal planes (see inset diagram). This section follows the same direction as the tractus olfactorius lateralis. On account of this inclination, I was able to observe axons of cells of the nucleus tractus olfactorii lateralis entering the stria terminalis. This could never be observed in transverse or sagittal sections owing to the course followed by this component of the stria.

Certain interesting peculiarities have been observed in this section. Several collaterals of mitral axons of the tractus olfactorius lateralis (T.Of.) enter the nucleus tractus olfactorii lateralis (N.T.Of., fibers *1* to *11*); fibers *1* and *2* approach cell *b*; fiber *3* courses in the same manner toward cell *c*; fiber *4*, provided with delicate sprouts, spreads out on one descending dendrite of cell *c*; fibers *6*, *7*, *8*, and one twig of fiber *9* form the nest *12*; fibers *10* and *11* form the nest *13*. From the nucleus tractus olfactorii lateralis (N.T.Of.) the cells *a*, *b*, *c*, *d*, and *e* send off their axons (*1a*, *1b*, *1c*, *1d*, and *1e* respectively) to enter the rostral part of the stria terminalis. It should be observed that several fibers coursing initially in the stria leave the bundle, bend caudally, and follow a sagittal course, perhaps to reach caudal levels of the amygdaloid complex (see fibers *16* to *20*).

The stria terminalis takes a caudomedial exit from the amygdaloid complex, coursing closely associated to the tail of the caudate nucleus. At commissural levels, the bundle bends and descends ventrally. A little above the transverse limb of the commissura anterior and near the midline it spreads out, forming different components that go toward various destinations.

Figure 53 is a sagittal section through the hypothalamus lateralis (H.L.) taken from a 6-day-old mouse. The different components of the stria terminalis are labeled according to Johnston (1923). The S.t.4 or precommissural component appears as a group of fibers curving around the commissura anterior and passing in a caudal direction among the innumerable sagittally running fibers that course throughout the hypothalamus lateralis. This component gives off collaterals to the gray matter in front of the commissura.

The postcommissural components S.t.2 and S.t.3 (Fig. 53) of the stria cannot be separated from each other in the mouse Golgi sections. The bundles spread out in the region underneath the commissura anterior and only a few fibers appear

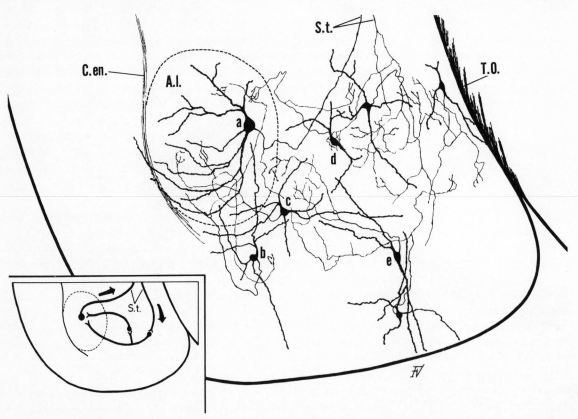

Fig. 52. Mouse 20, 20 days old. Transverse section through the amygdaloid complex. A thin amygdalopetal fiber of the stria terminalis (S.t.) enters the medial amygdaloid region. Cells *b*, *c*, *d*, and *e* send their axons toward the nucleus amygdaloideus lateralis (A.l.). In the latter, cell *a* projects into the stria. The inset diagram explains this circuitous pathway within the amygdaloid complex. Golgi method. (From Valverde, 1962.)

to continue caudally in the medial forebrain bundle (see Fig. 31, fibers *10*, *11*, and *12*). The majority give off many collaterals, forming the complicated system of nests in the region of the hypothalamus lateralis (H.L.) pierced longitudinally by some thick fibers coming from more rostral levels.

The S.t.1 or the commissural component, located between S.t.4 and S.t.3, is formed of fine fibers reaching the superior edge of the commissura (Fig. 53) which then turn sharply to cross the midline.

Figure 54 is a transverse section through the commissura anterior in a 9-day-old rat. This figure shows that the medial half of the stria terminalis (S.t.) enters to form the commissural component of the stria (S.t.com.) and that this component actually is of the same size as each of the other two components of the commissura: the pars tempo-

ralis (C.A.p.t.) and the pars bulbaris (C.A.p.b.). The lateral half of the stria terminalis arborizes in the bed nucleus of the stria terminalis (B.S.t.), crosses the commissura, and enters the regio praeoptica (R.P.O).

The commissural component of the stria could not be followed in Golgi sections to its termination. In experimental material I could trace it to the contralateral bed nuclei of the stria and the commissura anterior, but never to the contralateral claustrum or to the contralateral nucleus tractus olfactorii lateralis, as Sprenkel (1926) described. As shown in Fig. 32, there are several cells located in the proximity of the commissura anterior (cells *h*, *i*, *j*, *k*, and *l*) with one or two dendrites extending into the commissura; synaptic contacts can be expected to exist from this disposition.

The bed nucleus of the stria terminalis consists

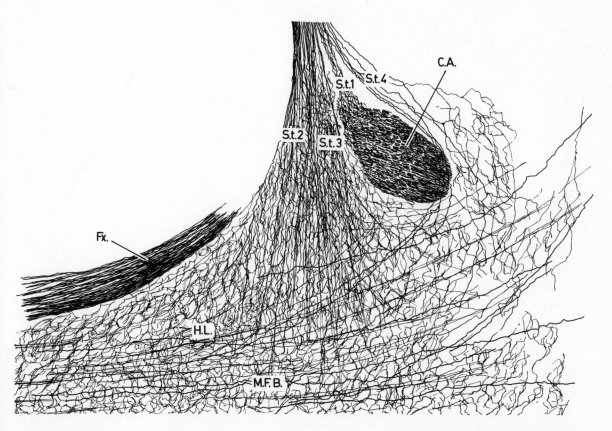

Fig. 53. Mouse 18, 6 days old. Sagittal section through the lateral hypothalamus, showing the termination and distribution of four of the Johnston's components of the stria terminalis. Golgi method. (From Valverde, 1963a, c.)

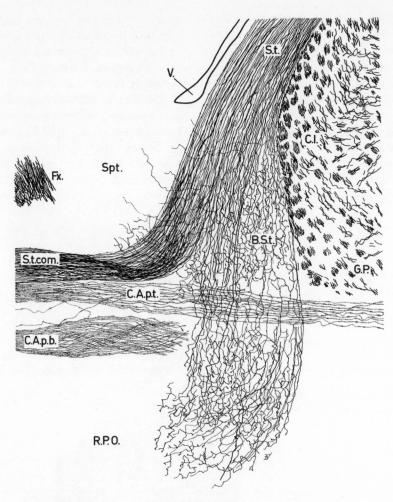

Fig. 54. Rat 43L, albino, 9 days old. Transverse section through the commissura anterior, showing the pars bulbaris (C.A.p.b.) and temporalis (C.A.p.t.) of that bundle, the commissural component of the stria terminalis (S.t.com.), and terminal ramifications of the stria in the bed nucleus (B.St.) and regio praeoptica (R.P.O.). Golgi method.

of a group of cells scattered throughout the stria terminalis; at commissural levels it develops into a relatively larger expanded cell area. Its position and connections represent one of the most interesting formations of the amygdala and related structures. According to Sprenkel (1926), the bed nucleus represents a direct continuation of the nucleus amygdaloideus centralis. It becomes continuous with the bed nucleus of the commissura anterior and the bed nucleus of the medial forebrain bundle. Münzer and Wiener (1902) included the bed nucleus of the stria terminalis at commissural levels

within the anteromedial nucleus of the thalamus.

Figure 55 is a sagittal section of the expanded portion of the bed nucleus in a 6-day-old mouse. In this figure it can be seen that the axons of its cells behave in one of three different manners: there are cells (a, c, d, e, f, and g) whose axons divide into two branches, coursing in opposite directions, one of which follows the stria terminalis (S.t.) while the other passes to the regio praeoptica; a second category of cells is represented by b, h, i, j, and k possessing axons directed only toward the regio praeoptica; the third class is represented by cell *l*, whose axon courses in the stria terminalis as a single fiber without emitting collaterals.

The bed nucleus of the stria terminalis represents a source of fibers coursing toward the amygdaloid complex. This fact was early suggested by Déjerine (1901); from observations in the rat, Gurdjian (1928) also suggested that in this animal the stria terminalis consists of afferent and efferent fibers; Lauer (1945) arrived at the same conclusion. Nauta (1958) found, after lesions in the preoptic area in the cat, some degenerated fibers running in the stria terminalis. In the rat, Shute

and Lewis (1961), using the modification of the Koelle cholinesterase technique (Lewis, 1961), traced fibers, originated in the bed nucleus of the stria terminalis, into the amygdaloid complex via the stria terminalis, and Powell, Cowan, and Raisman (1963) also traced degenerating stria terminalis fibers into the amygdala after lesions in the hypothalamic and preoptic areas in the rat. The studies of Hilton and Zbrożyna (1963) support this fact.

In my Golgi sections terminal fibers were systematically observed reaching the amygdala via the stria terminalis. The fibers are located in the

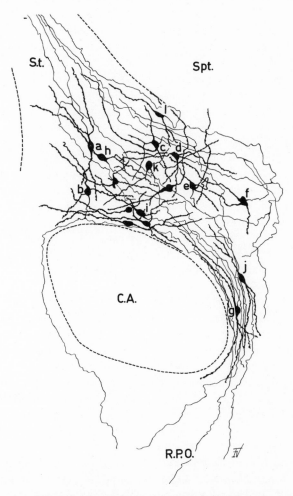

Fig. 55. Mouse 21, 6 days old. Sagittal section through the commissura anterior, showing several cells in the bed nucleus of the stria terminalis. Golgi method. (From Valverde, 1963a.)

medial part of this bundle and they ramify in the central and medial nuclei connecting with cells placed here, whose axons would either link with other lateral amygdaloid cells (Fig. 52) or enter again as amygdalofugal stria fibers. In Fig. 49 the amygdalopetal stria fibers have been drawn in red. As mentioned previously, the dendrites of cells of the nucleus amygdaloideus centralis (A.c.), oriented in the transverse plane, appear to receive these afferent impulses.

In my experimental series I traced a system of fibers, apparently of thalamic origin, to rostral thalamic levels where some terminate in the bed nucleus of the stria terminalis, others, few in number, curve dorsally, enter the stria terminalis, and run toward the amygdala (see Cat 15). The existence of such fibers was verified in Golgi sections of the albino rat and, together with certain interesting peculiarities of the structure of the bed nucleus of the stria terminalis, are shown in Fig. 56. This figure is a sagittal section of a 4-day-old rat in which the cells belonging to the bed nucleus have been drawn. It should be observed that the nucleus is really large in this animal. It is pierced sagittally by a great number of fibers of the rostral thalamo-cortical radiation. This radiation, as it passes from the thalamus (Thal.) to the nucleus caudatus-putamen (Cd.Put.) through the narrow passage existing between the fornix (Fx.) and the commissura anterior (C.A.), becomes compact and makes numerous synaptic contacts with the dendrites and bodies of cells of the bed nucleus. The orientation of many dendrites perpendicular to the thalamo-cortical radiation is impressive, as is also the great number of dense pericellular nests formed by collateral twigs of the thalamo-cortical radiation in the bed nucleus; several nests have been drawn (1 to 9); nest 9 is formed by the final ramification of an ascending fiber.

It cannot be disregarded that fibers piercing the bed nucleus, at least in part, represent cortical projection fibers, but it should be pointed out that in the rat and the cat the bed nucleus represents a synaptic relay between the thalamus and the amygdaloid complex, probably related to transmission toward the latter of impulses traveling through the diffuse thalamic projection system.

In Fig. 56 cell a sends its axon (1a) to the stria terminalis (S.t.); cells b and i possess axons (1b and 1i respectively) directed rostrally, but I could not ascertain whether or not cell b belongs to the bed nucleus or to the adjoining reticular nucleus of the thalamus (R.) with which the former becomes continuous and fused in some places. The axon 1c of cell c gives off the collateral 2c, which could be followed entering the stria terminalis; two collaterals (2d and 3d) of the axon 1d of cell d also were followed to the stria terminalis; the axon 1e of cell e behaves in the same manner and the axons 1f, 1g, and 1h belonging to cells f, g, and h could be followed for a distance sufficient to suggest their entrance into the stria terminalis.

A small bundle composed of fine fibers (15), presumably of thalamic origin, bends dorsally, entering the stria terminalis. The thick fibers labeled 10, 11, and 12 behave in the same fashion. Fibers 13 and 14, coursing in the opposite direction, are collaterals of fibers 10 and 11.

I gained the impression that the interchange of fibers between the stria terminalis and the stria medullaris is uncertain; I could never trace degenerating elements from the stria terminalis into the stria medullaris (the so-called S.t.5 or stria medullaris component of the stria terminalis).

Besides the direct entrance, just described, of fibers of probable thalamic origin into the stria terminalis, collateral fibers of sagittally running ones also were observed (see collaterals 16, 17, and 18 in Fig. 56).

The fasciculus longitudinalis associationis

This system, connecting the amygdala with the area amygdaloidea anterior and lateral parts of the regio praeoptica through the ventral part of the brain, has been partly described previously when I dealt with the cellular structure of the area amygdaloidea anterior (see p. 68, and Valverde, 1962, 1963a, b, c). It forms the compact part of the rostral projection system of the amygdala.

In the cat this system (see Fig. 57, F.l.a.) forms a well-circumscribed bundle originating mainly in the basolateral group of the amygdala and, following in a rostromedial direction, it spreads out in the region underneath the temporal limb of the commissura anterior. It has wide connections with

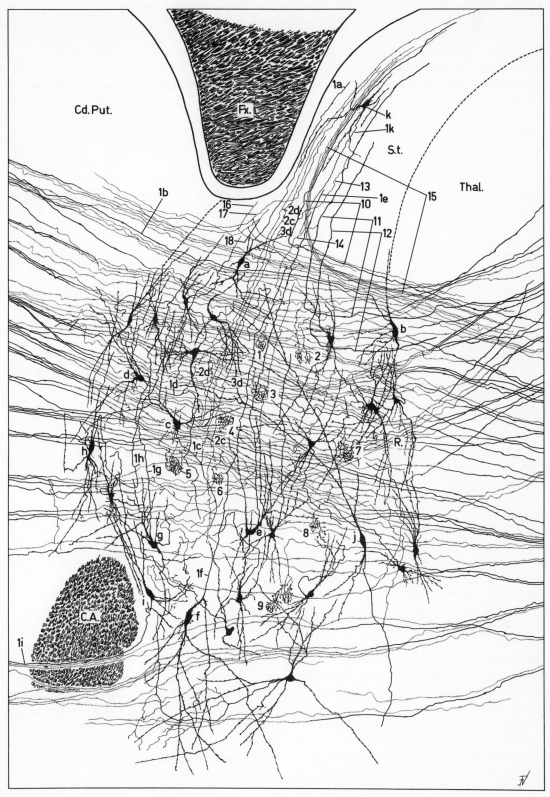

Fig. 56. Rat 40R, albino, 4 days old. Sagittal section through the bed nucleus of the stria terminalis. Golgi method.

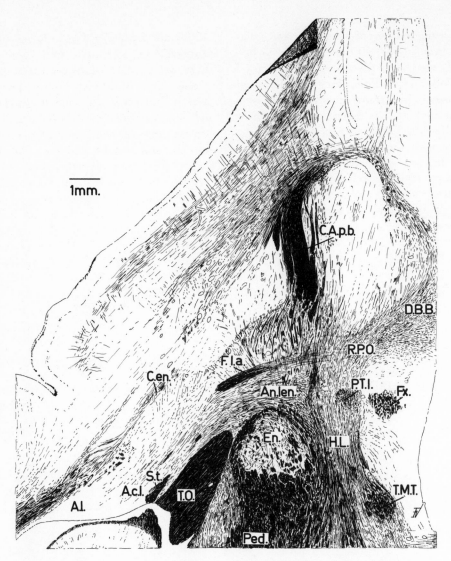

Fig. 57. Horizontal section through the ventral part of the forebrain of an adult cat. Heidenhain stain.

the area amygdaloidea anterior and, farther forward, with the area praepiriformis, the lateral preoptic and hypothalamic regions, the tuberculum olfactorium, and the regio limbica anterior.

In Heidenhain horizontal sections of the cat's brain, the fasciculus longitudinalis associationis (Fig. 57, F.l.a.) forms a well-circumscribed system of fibers which can be traced along the rostral edge of the ansa lenticularis (An.len.) to the regio praeoptica (R.P.O.), where it fuses partly with the diagonal band of Broca (D.B.B.).

In the mouse and rat this system is formed by cells with short axons in which the area amygdaloidea anterior with its cells forms a part. In these animals no long fibers were observed in this region; its origin is located in the amygdala in the regions where the diffuse interconnections of the intermediate-axoned type of cell form its substratum.

The group of amygdaloid cells (Amyg.) drawn medial to the commissura anterior, pars temporalis (C.A.p.t.) in Fig. 29 and the cells of the area amygdaloidea anterior *f* to *u* of Fig. 50 illustrate the variable destinations of their axons and collaterals; however, in Fig. 50 there appears to be a preferential rostral orientation of their axons.

In phylogenetic evolution the fasciculus longitudinalis associationis shows notable differences: in the mouse and rat it is represented by a short-axoned cell system connecting the amygdala with the area amygdaloidea anterior and the regio praeoptica; in the cat it is formed by compact bundles running in the region below the temporal limb of the commissura anterior and connecting the amygdala with the regio praeoptica, anterior amygdaloid and praepiriform areae, and diagonal band of Broca, and the regio limbica anterior (Johnston, 1923; Fox, 1940; Lammers and Lohman, 1957; Valverde, 1962, 1963a, b, c); in the monkey it forms a rather compact system emerging from the amygdala and spreading medially in the region underneath the lentiform nucleus, distributing fibers to certain basal forebrain structures (Nauta,

1961), and finally, in man, it forms an important amygdalo-hypothalamic and amygdalo-septal fiber system which passes ventral to the lentiform nucleus and the internal capsule, forming a sizable fiber tract that is actually larger than the stria terminalis and of about the same size as the commissura anterior (Klingler and Gloor, 1960). Therefore, it should be observed that, while the fasciculus longitudinalis associationis gains in size as it ascendes in the phylogenetic scale, parallel observations suggest that the opposite occurs with the stria terminalis, which is a sizable bundle in the mouse and can be considered as a vestigial or regressive system in man. This reciprocal evolution develops probably as a consequence of the increase of the basolateral amygdaloid complex and of the area amygdaloidea anterior over the progressive involution of the corticomedial part and their related preoptic and hypothalamic regions through the stria terminalis.

4   The Connections of the Amygdaloid Complex. General Discussion.

(I) EFFERENT AMYGDALOID CONNECTIONS

(1) *The amygdaloid component of the stria terminalis*

In my experimental series, the cases yielding amygdalofugal degeneration throughout the stria terminalis correspond to Cats 3, 5, 6, and 24 (see Figs. 5, 7, 8, 9, 10, and 24). In Cats 3, 6, and 24 the commissural component of the stria was degenerated; in these cases their corresponding lesions, although different in size, occupy a similar position (see Fig. 3, sections 11 and 12). Fine degenerating fibers accompanying this component before they enter the commissura disperse among the cells of the bed nucleus of the stria terminalis. In support of my Golgi observations, it was observed that the commissural component synapses with dendrites of the surrounding cells of the bed nucleus of the commissura anterior. Since the commissural component could not be traced to any contralateral structure and is apparently related exclusively to the two bed nuclei of the stria terminalis and the commissura anterior, it should be reasonable to label it *bed nuclei component of the stria terminalis.*

According to Johnston (1923), this bundle represents a commissural system between the nuclei tractus olfactorii laterals of the two sides. Sprenkel (1926) assumed also that this bundle links these two nuclei as well as the contralateral claustrum. Humphrey (1936) and Fox (1940, 1943) substantiated the origin of this commissural component in the nucleus tractus olfactorii lateralis and Fukuchi (1952) supports the view that, besides this origin, contributions from the medialis and centralis amygdaloid nuclei also join this component.

Regarding the termination of the commissural

bundle, the experimental observations show marked differences with respect to those made on normal material; Fox (1943), Lammers and Magnus (1955), Lammers and Lohman (1957), Omukai (1958), Ban and Omukai (1959), and Nauta (1961) could never follow this bundle beyond the neighborhood of the commissura anterior. This is in accordance with my experimental observations.

As mentioned previously, the commissural bundle relates the amygdala with the two bed nuclei of the stria terminalis and the commissura anterior and, therefore, it cannot be considered as a commissural system, a point of view already supported by Lammers and Magnus (1955). However, it should be pointed out that in the rat this part of the stria terminalis forms a sizable bundle (see Fig. 54) that probably synapses with other cell bodies and dendrites besides the dendrites protruding into the commissura. In my Golgi sections I could never follow it contralaterally, but it can be assumed that in this animal the bundle might be related to the contralateral bed nucleus of the stria terminalis which in this animal forms a well-developed nuclear mass.

The lesion of Cat 5 did not yield degeneration in the commissural bundle, but the septal and preoptic components of the stria were degenerated. The septal component (bundle 4 of Johnston) appears, however, to be of minimal volume and, like the preoptic (postcommissural or bundles 2 and 3 of Johnston), it overlaps widely with the degeneration traced via the fasciculus longitudinalis associationis. Neither component was observed by Lammers and Magnus (1955) in the cat.

Johnston (1923) pointed out that the basal nucleus receives (this author does not describe the direction of conduction) important bundles from the stria terminalis as well as, although in a lesser degree, from the lateral nucleus. Sprenkel (1926) states that part of the infra- and supracommissural bundles of the stria are connected with the lateral part of the amygdala, which includes the basal, accessory basal, and lateral nuclei. Recent studies of Hall (1963) in the cat show that the basal nucleus contributes to the supracommissural component of the stria terminalis.

In the cat, Fox (1943) failed to trace degeneration in the stria after lesions in the lateral nucleus of the amygdala. He demonstrated that lesions in the caudomedial part of the amygdala show degeneration in the supracommissural and preoptic components of the stria.

It has been assumed that the stria terminalis arises largely in the corticomedial complex, including the nucleus tractus olfactorii lateralis, and in part of the basal nucleus (Young, 1936; Kappers, Huber, and Crosby, 1936; Jeserich, 1945; Lauer, 1945; Klingler and Gloor, 1960; Hall, 1963). Adey and Meyer (1952) and Nauta (1961) stated that in the monkey the stria terminalis originates largely in the caudal half of the amygdaloid complex.

The experiments of Gloor (1955a, b) in the cat show the intimate connection of the corticomedial complex with the stria terminalis. Stimulation within this complex produced short-latency responses in the stria terminalis, while the stimulation of the basolateral complex produced long-latency responses in this tract. This is interpreted by Gloor as confirmation of previous anatomic studies, and he points out that in the cat the stria terminalis is fed by the neurons of the corticomedial complex and of the posterior part of the basal nucleus while the remainder of the amygdala contributes only indirectly to the stria terminalis.

Contrary to the observations of Nauta (1961) in the monkey, my experimental studies in the cat suggest that substantial contributions of the stria terminalis and, more specifically, those related to the septal and postcommissural components arise in the rostral parts of the amygdala (Cat 5), while the commissural component originates in the cau-

domedial parts of the amygdala (Cats 3, 6, and 24), including both the medialis and lateralis subdivisions of the nucleus amygdaloideus centralis.

Much disagreement appears to exist concerning the termination of the postcommissural component of the stria terminalis. Cajal (1911) traced fibers of the stria terminalis through the hypothalamus to the subthalamic region. Johnston (1923) and Sprenkel (1926) suggested that these fibers could be traced caudally in the medial forebrain bundle. Kappers, Huber, and Crosby (1936) have traced stria terminalis fibers to the premammillary nuclei. Adey and Meyer (1952) in the monkey traced stria terminalis fibers bilaterally to the nucleus hypothalamicus ventromedialis; while Nauta (1961) in the same animal could not follow such fibers caudal to chiasmatic levels.

In my experimental series the degeneration traced through the hypothalamus lateralis in Cat 5, dorsal and ventral with respect to the hypothalamic fornix, to the nucleus subthalamicus, although supported in part by previous investigations of other workers, seems to me far less than conclusive. It could not be traced in continuity with the stria terminalis and, as mentioned in the description of this experiment, this hypothalamic degeneration might represent direct, medial, amygdalo-hypothalamic pathways.

In the mouse and the rat the hypothalamic connections of the amygdala via the stria terminalis appear to be more massive than in the cat and the monkey, and thus, as mentioned previously (see pp. 98–99), the postcommissural components, identified as bundles 2 and 3 of Johnston (1923), spread out behind the commissura anterior, contributing largely to build up the complicated network of fibers of the hypothalamus.

The relations established between the stria terminalis and the medial forebrain bundle have been discussed previously (see p. 72).

Septal contributions of the stria terminalis were not observed by Lammers and Magnus (1955) in the cat, although these authors showed the existence of degenerating fibers passing just rostral to and around the commissura anterior to the nucleus accumbens septi. These fibers appear to be similar to those traced in Cat 5 to this nucleus, even though many contributions enter the adjoining septal regions. Nauta (1961) failed to observe amygdalo-septal fibers in the monkey via the stria terminalis.

Both my experimental and my Golgi studies have failed to confirm the existence of a stria medullaris component of the stria terminalis. In a previous study (Valverde, 1963a) I observed in frontal Golgi sections of the mouse some fibers that, at commissural levels, curve sharply dorsally and join the fornix, but the assumption that these fibers enter the stria medullaris could not be supported.

The existence of the stria medullaris component has been described in normal material by some authors (Honegger, 1892; Johnston, 1923; Gurdjian, 1925; Sprenkel, 1926; Hilpert, 1928; Humphrey, 1936; Young, 1936; Lauer, 1945; Fukuchi, 1952; Thomalske, Klingler, and Woringer, 1957; Klingler and Gloor, 1960). Bürgi (1954), after damage to the stria medullaris on one side, traced degenerating fibers caudally in this stria to cross in the commissura habenularum to the contralateral stria medullaris, where the fibers run rostrally to join the stria terminalis and thence backward to the amygdala. Similar observations have been reported by Mitchell (1963). Cragg (1961a) in the rabbit, after lesions in the amygdaloid region, observed fibers passing from the stria terminalis to the stria medullaris.

Other authors employing experimental methods (Fox, 1943; Adey and Meyer, 1952; Lammers and Magnus, 1955; Nauta, 1961) were unable to trace this component.

The degeneration traced in Cat 11 through the stria medullaris to the lateral and medial habenular nuclei, although rather scanty, appears to originate in the base of the septum and dorsal parts of the regio praeoptica. These fibers were observed early by Vogt (1898) in the mouse and, with minor variances, recently confirmed experimentally by Nauta (1956), Fortuyn, Hiddema, and Sanders-Woudstra (1960) and Powell (1963) in the rat; Nauta (1958) and Guillery (1959) in the cat, and Cragg (1961a) in the rabbit.

Amygdalofugal components of the stria terminalis joining the fornix have been described by Honegger (1892) and Kölliker (1896) and recently

by Klingler and Gloor (1960). The existence of these fibers was denied by Cajal (1911).

(2) Ventral amygdalofugal pathways

(a) Rostral projection system of the amygdala. Under this name I have included two groups of fibers (lateral and medial) which, originating in the amygdala, course rostrally to the area praepiriformis and the regio praeoptica and farther forward to the diagonal band of Broca and the regio limbica anterior.

The lateral part of this system is composed of diffuse and relatively short fibers extending from the amygdala to the area amygdaloidea anterior and the area praepiriformis. This system is clearly represented in Cats 1, 3, 4, 5, 6, and 24 (see Figs. 4, 5, 6, 7, 8, 10, and 24) by the diffuse degenerating fibers that occupy the area amygdaloidea anterior (A.a.), lateral to the fasciculus longitudinalis associationis (F.l.a.) and in the depth of the area praepiriformis (A.pp.).

It should be noted that the more rostral lesions (Cats 1 and 5) give more abundant degeneration in the area amygdaloidea anterior and area praepiriformis, suggesting that this diffuse system originates mainly in the rostral amygdala.

The medial or compact part of the rostral projection system of the amygdala is the fasciculus longitudinalis associationis. A clear picture of its termination and distribution has been gained from my experimental observations. The animals showing degeneration of the fasciculus longitudinalis associationis correspond to Cats 1, 3, 4, 5, 6, 11, and 24 (see Figs. 4, 5, 6, 7, 8, 9, 10, 13, and 24). The degenerating fibers that made up the bulk of this fasciculus were constantly found ventral to the nucleus ansae lenticularis and lying dorsal to the limits separating the regio praeoptica from the area amygdaloidea anterior. The exact site of termination is located in the dorsal part of the lateral half of the regio praeoptica where it overlaps widely with the preoptic component of the stria terminalis.

More rostrally the fasciculus longitudinalis associationis can be traced to the diagonal band of Broca and the regio limbica anterior (Cats 1, 3, 5, and 11; see Figs. 4, 5, 9, and 13). Lesions causing this degeneration are the rostral ones of the group, indicating that these rostral terminations originate in the rostral amygdala. Further information was obtained from Cat 11, in which the lesion in the regio praeoptica suggests that the connections between the amygdala and diagonal band of Broca and base of the septum (see Fig. 13, section 2) via the fasciculus longitudinalis associationis relay largely in or around the zone destroyed by the coagulation. From here the rostrodorsal continuation of the fasciculus longitudinalis associationis toward the base of the septum is formed by ascending fibers fused with the Zuckerkandl's radiation, which, with the diagonal band of Broca, should be considered as a rostral continuation of the fasciculus longitudinalis associationis.

Concerning the origin of the fasciculus longitudinalis associationis within the amygdala, little information has been gained; however, it appears that this fasciculus originates mainly in the basolateral complex of the amygdala. This latter assumption would agree with the observations reported by Fox (1943), Lammers and Magnus (1955), and Nauta (1961).

Johnston (1923) and Fox (1943) suggested that fibers of the fasciculus longitudinalis associationis join the medial forebrain bundle. This is in disagreement with my observations, since, as will be discussed below, the amygdalo-hypothalamic connections relating the amygdala with the medial forebrain bundle take a more caudal exit from the amygdala and follow a sublenticular transit through the ansa lenticularis.

My observations appear to confirm early suggestions of Johnston (1923), who considered that the fasciculus longitudinalis associationis unites rostrally with the diagonal band of Broca.

Lammers and Lohman (1957), Nauta and Valenstein (1958), and Nauta (1961) traced fibers of the rostral-projection system of the amygdala to the rostral limbic cortex; their distribution resembles closely that which I traced to the regio limbica anterior.

Nauta and Valenstein (1958) and Nauta (1962) have traced fibers of the fasciculus longitudinalis associationis to the tuberculum olfactorium. The existence of fibers of this fasciculus ending in the bed nucleus of the stria terminalis as reported by Fox (1943) seems to me uncertain. I have not been

able to trace such connections in either Golgi or experimental material. Conversely, the termination of the fasciculus in the regio praeoptica agrees well with the observations reported by Fox (1940), Lammers and Magnus (1955), Lammers and Lohman (1957), Hall (1960, 1963), and Nauta (1961).

Further details of the fasciculus longitudinalis associationis and the short-linked system of neurons which, in the mouse and rat, represent this fasciculus have been discussed already (see pp. 68–69 and 102–105).

(b) *The medial amygdalo-hypothalamic pathway.* The cases showing degeneration of this system correspond to Cats 1, 3, 5, 6, and 24 (see Figs. 4, 5, 7, 8, 9, 10, and 24). This pathway has been traced rostromedially from the amygdala, through the ansa lenticularis, to the hypothalamus lateralis, where it links with the medial forebrain bundle. It appears to originate largely in the nucleus amygdaloideus centralis, pars medialis, as well as in the area amygdaloidea anterior.

As shown by Cat 6 (see Fig. 10, section 2), the fibers condense in a fascicle located in the medial part of the medial subdivision of the nucleus amygdaloideus centralis. The abundant fiber degeneration observed in Cat 1 in this system suggests that it increases in its rostromedial course through the medial subdivision of the nucleus amygdaloideus centralis either by contributions of this nucleus or by adding fibers from the area amygdaloidea anterior which envelops the rostral parts of the former.

The amygdalo-hypothalamic pathway here described would correspond to the hypothalamic component of the ventral amygdalofugal system traced by Lammers and Magnus (1955), Lammers and Lohman (1957), Hall (1960, 1963), and Nauta (1961), although the different course followed by this amygdalo-hypothalamic pathway justifies a separate description.

The medial amygdalo-hypothalamic pathway becomes closely associated with the temporo-hypothalamic projections traced in Cat 17 (see Fig. 20). The latter projection joins the former as it passes through the sublenticular region forming the ansa lenticularis, which represents the pathway connecting the temporo-amygdaloid complex with extensive regions of the hypothalamus and mesencephalon (see Fig. 20).

The above connections should not be confused with the direct and indirect pathways relating the temporo-amygdaloid complex with the dorsal thalamus. These latter fibers course rostral to the ansa lenticularis through the pedunculus thalami extracapsularis of Ludwig and Klingler (1956) and will be discussed below.

(3) *Amygdalo-hippocampal connections* (Fig. 58)

The existence of fibers connecting the amygdala with the hippocampus, although described by several authors (Johnston, 1923; Gurdjian, 1928; Hilpert, 1928; Loo, 1931; Mittelstrass, 1937; Fukuchi, 1952), have been denied by others (Al-

len, 1948; Pribram, Lennox, and Dunsmore, 1950; Adey and Meyer, 1952; Nauta, 1961). The experiments of Gloor (1955a) have confirmed the absence of direct connections between both structures.

The amygdala appears to send fibers to the cortex piriformis (Johnston, 1923; Gurdjian, 1928; Hilpert, 1928; Mittelstrass, 1937; Crosby and Humphrey, 1944; Lauer, 1945; Gloor, 1955a), and therefore from the prepiriform and periamygdaloid cortices caudalward, impulses could be conducted to the area entorhinalis through a series of short chains, as has been described previously (see pp. 85–86). The existence of prepiriform-entorhinal fibers appears to have been observed by Cajal (1901), who labeled this system

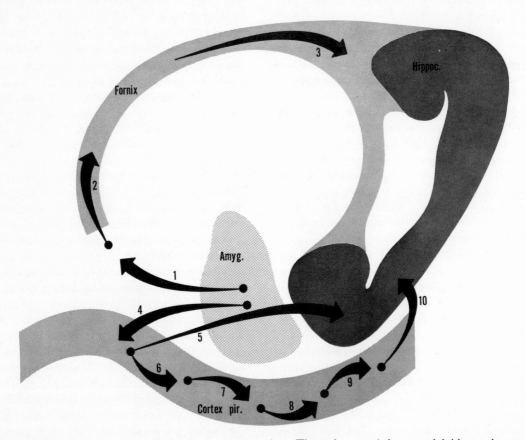

Fig. 58. Schema of the amygdalo-hippocampal connections. Through arrow 4 the amygdaloid complex projects to the cortex piriformis; from the latter, impulses reach the hippocampus, either directly (arrow 5) or via the multisynaptic pathways of the cortex (arrows 6, 7, 8, 9, and 10). Through arrow 1 (ventral amygdalofugal pathways) impulses reach the septal and preoptic regions, which in turn project via the fornix (arrows 2 and 3) to the hippocampus.

voie sagittale d'association de l'écorce temporale (Cajal, 1911)[1]. From the entorhinal cortex, the alvear and perforant bundles described by Cajal (1911) and Lorente de Nó (1934) enter the hippocampus and form the last link in this multisynaptic amygdalo-piriform-entorhino-hippocampal pathway.

The conduction of impulses through the cortex piriformis has been extensively studied by several workers (Petr, Holden, and Jirout, 1949; Pribram, Lennox, and Dunsmore, 1950; Kaada, 1951; Berry, Hagamen, and Hinsey, 1952; Hugelin *et al.*, 1952; Adey and Meyer, 1952; MacLean and Pribram, 1953; Blackstad, 1956; Lammers and Lohman, 1957; Cragg, 1961a, b). Their existence points to the wide sphere of influence of the cortex piriformis on several cortical and subcortical centers in which the hippocampus forms a part. It was generally assumed that the piriform-hippocampal conduction is composed of multisynaptic chains, but the findings reported in Cat 5 of my experimental series, confirming previous observations of Cragg (1961b), indicate that, besides the multisynaptic conduction reported previously, direct prepiriform-hippocampal connections do exist.

Of the three systems of fibers described in Cat 5 (see pp. 22–23) under paragraph (*f*) the groups *a* and *b* (Fig. 7) would represent axons of piriform cells, interrupted by the lesion, traceable dorsocaudally through the nucleus amygdaloideus lateralis toward the hippocampus. These fibers course through the systems described as relating the cortex piriformis with the lateral parts of the amygdaloid complex (see pp. 86–94). System *c* of Cat 5 has been interpreted as being formed by the ascending axons and collaterals of certain cells of the cortex piriformis entering the fibrillar layer of this cortex.

The fact that only in this experiment were fibers traced to the hippocampus can be explained on the basis that these piriform-hippocampal fibers course through the nucleus amygdaloideus lateralis and through the capsula externa, regions not involved in the rest of the present experiments. On the other hand, the negative findings in Cat 4 concerning piriform-hippocampal fibers, in spite of the proximity of its lesion with respect to that of Cat 5, lend support to the assumption

that the piriform-hippocampal fibers originate rostrally in the prepiriform area and presumably in the area amygdaloidea anterior (where direct olfactory fibers end) and not in the periamygdaloid cortex. This would explain the negative findings reported by Adey and Meyer (1952).

According to Cragg (1961b), in the cat direct olfactory bulb fibers end more rostrally in the piriform lobe than in the rat and, consequently, the piriform-hippocampal direct connections in the latter originate more caudally.

It should be noted that, contrary to the massive and well-established piriform-amygdaloid relations, evidence of amygdalo-piriform connections have rarely been observed either in my experimental series or in my Golgi sections (see p. 94). In Cat 5 (see Figs. 7 and 8) diffuse fibers, not belonging to the well-defined systems *a*, *b*, and *c* traced in this animal, could be followed caudally entering the nucleus amygdaloideus corticalis and, sparingly, the adjoining periamygdaloid cortex. In Cats 1 and 24 (see Figs. 4 and 24) a few degenerating fibers can be seen piercing the capsula externa entering the depth of the periamygdaloid cortex.

Gloor (1955a) suggested that amygdalo-hippocampal connections may take place via the multisynaptic chains of the piriform lobe to the hippocampus (Fig. 58, arrows 4, 6, 7, 8, 9, and 10); however, on the basis of previous investigations of other workers and my own observations, it becomes conceivable to suggest that, besides the conduction through the multisynaptic chains of the cortex piriformis, indirect amygdalo-hippocampal connections (Valverde, 1964c) might occur in the two following manners:

(*a*) Through the rostral projection system (Fig. 58, arrow 4), the amygdala would link with those regions of the cortex piriformis and area amygdaloidea anterior that give rise to the direct piriform-hippocampal system of fibers (Fig. 58, arrow 5) observed in Cat 5. This is supported by the findings reported by Cragg (1961b).

(*b*) Throughout the rostral projection system of

---

[1]Dorsoventral associative pathway of the sphenoidal cortex (Cajal, 1955).

the amygdala (Fig. 58, arrow 1) to the regio praeoptica, nucleus of the diagonal band of Broca and septum, and thence backward, through the fornix (Fig. 58, arrows 2 and 3) to the hippocampus, as reported by Crosby (1917), Rose and Woolsey (1943), Morin (1950), Daitz and Powell (1954), Green and Adey (1956), Cragg and Hamlyn (1957), Votaw and Lauer (1963), Powell (1963), and incidentally observed in Cat 11 (see Figs. 13 and 14) of my experimental series.

Through the fornix system, the hippocampus projects to the hypothalamus lateralis and medial forebrain bundle area (Lauer, 1945; Sprague and Meyer, 1950; Crosby and Woodburne, 1951; Simpson, 1952; Powell and Cowan, 1955; Guillery, 1956). These connections are established either directly (cat) or indirectly (rat and guinea pig) with relays in the septum and regio praeoptica (Nauta, 1956; Valenstein and Nauta, 1959). In Cat 6 (see Figs. 10 and 11) of my experimental series, the lesion encroached widely upon the ventral part of the hippocampus; degenerating fibers could be traced through the fornix to the hypothalamus lateralis and medial forebrain bundle area but not to the mammillary nuclei, which confirms the observations referred to above. These hippocampo-hypothalamic fibers traced in Cat 6 originate in the region where the piriform-hippocampal fibers traced in Cat 5 end, that is, in the ventral hippocampus.

Hence, we can consider that these regions of the hypothalamus lateralis and medial forebrain bundle area covered by the projections originated in the ventral hippocampus represent the hypothalamic projection area of a transhippocampal amygdalo-hypothalamic pathway. It should be observed, also, that this hypothalamic region overlaps widely with the zone of distribution of the direct amygdalo-hypothalamic fibers traced previously (see pp. 109–110).

The amygdala appears thus widely interconnected with the medial forebrain bundle area through at least four pathways: through the postcommissural component of the stria terminalis; through the fasciculus longitudinalis associationis; through the direct medial amygdalo-hypothalamic fibers; and, finally, indirectly through the amygdalo-piriform-hippocampal-hypothalamic pathway just described.

It has been suggested (Gastaut and Lammers, 1961) that the amygdalo-mesencephalic connection observed by Gloor (1955a) is mediated through the entorhino-mesencephalic fibers described by Adey, Merrillees, and Sunderland (1956).

(4) Amygdalo-temporal connections

Evidence of amygdalo-temporal connections has been obtained in Cats 3, 6, and 24 (see Figs. 5, 10, and 24). As described previously (Valverde, 1963b), these amygdalo-temporal fibers course transversally through the dorsal limits of the amygdala underneath the putamen, taking a lateral exit from the amygdala, then pierce perpendicularly the capsula externa, and, traveling through the ventral parts of the claustrum, spread out in the white matter in front of the amygdala. The site of termination is the ventral region of the gyrus sylvianus posterior which, according to the homologies established by Gastaut and Lammers (1961), would correspond to the posterior part of the insula and to the tip of the temporal lobe of primates. In this respect, this connection is similar to the fasciculi amygdalo-temporalis and amygdalo-insularis described in man by Klingler (1940), Ludwig and Klingler (1956), and Klingler and Gloor (1960) and to the amygdaloid projections to the rostral parts of the superior, middle, and inferior temporal gyri and ventral insula traced by Nauta (1961) in the monkey.

No information has been gained from my experimental observations concerning the exact site of origin within the amygdala of this projection and yet it should be borne in mind that part of the amygdalo-temporal connections traced in this work might represent corticopetal fibers traveling through the ansa lenticularis interrupted either by the needle tract or by the area of coagulation.

Electrophysiologic evidence of amygdalo-temporal connections has been obtained by Pribram, Lennox, and Dunsmore (1950), Pribram and MacLean (1953), Segundo, Naquet, and Arana (1955), and Gloor (1955a). It is thought that the conduction takes place in both directions, although the temporo-amygdaloid direction of conduction appears to be more massive than the opposite one (Whitlock and Nauta, 1956; Nauta, 1961).

(5) *Amygdalo-thalamic connections*

Although Déjerine (1901) described the existence of fibers of the stria terminalis entering the stratum zonale of the thalamus, it appears that the first person to suggest the existence of amygdalo-thalamic connections was Hilpert (1928), though he limited his account to saying that probably the stria terminalis, along its course through the thalamo-caudate sulcus, might receive or send fibers from or to the thalamus.

Later, Fox (1949) described in the monkey degenerating Marchi fibers running in the inferior thalamic peduncle from the amygdala to the lateralis posterior, pulvinar, and medialis dorsalis thalamic nuclei. This amygdalo-thalamic connection has been confirmed by Nauta and Valenstein (1958) and Nauta (1961). As described by the latter author, the fibers relating the amygdala with the thalamus course toward the substantia innominata, where they enter the inferior thalamic peduncle and later the lamina medullaris medialis, continuing to the magnocellular part of the nucleus medialis dorsalis of the thalamus, where they terminate. Direct amygdalo-thalamic fibers were not observed by Hall (1963) in the cat after lesions in the basal and lateral amygdaloid nuclei.

Fortuyn, Hiddema, and Sanders-Woudstra (1960) have found that fibers arising in the ventral parts of the brain of the rat in a zone located between the tuberculum olfactorium and the nucleus amygdaloideus corticalis ascend parallel to the stria medullaris, as a component of the inferior thalamic peduncle, to the parataenial and medialis dorsalis thalamic nuclei. Powell, Cowan, and Raisman (1963) have traced in the rat fibers from the cortex piriformis to the nucleus medialis dorsalis of the thalamus.

These fibers appear to be similar to those described by Rioch (1931), Nauta (1958), and Guillery (1959) in the cat and by Cragg (1961a) in the rabbit. After coagulation of the substantia in-

nominata underneath the temporal limb of the commissura anterior, Escolar (1954) traced in the cat degenerating fibers through the ansa lenticularis; from here a dorsal stream curves around the medial border of the capsula interna, taking an ascending course toward the thalamus.

In Cat 5 (see Figs. 7, 8, and 9) of my experimental series abundant fiber degeneration was traced from the area of coagulation dorsally to enter the sublenticular stratum and via the inferior thalamic peduncle to terminate in the caudal parts of the nucleus medialis dorsalis of the thalamus. I assumed that these fibers originated in the rostral parts of the amygdala and area amygdaloidea anterior (Valverde, 1962, 1963b); later, I was inclined to believe that these fibers took origin in the area amygdaloidea anterior and that few, if any, contributions would be expected to come from the amygdala; however, as the field covered by the diverse lesions made in the present work did not fill the entire area of the amygdala (a considerable extension of the nucleus amygdaloideus basalis remained unexplored), I cannot give a definite answer to this problem; moreover, the direct amygdalo-thalamic connections traced by Nauta (1961) in the monkey originate largely in the basal nucleus. Electrophysiologic studies have confirmed the existence of connections between the amygdala and the temporal cortex from one side with the nucleus medialis dorsalis of the thalamus and adjacent midline cellular groups from the other side (Ajmone-Marsan and Stoll, 1951; Stoll, Ajmone-Marsan, and Jasper, 1951; Jasper and Ajmone-Marsan, 1952; Kaada, 1951, 1954; French, Hernández-Peón, and Livingston, 1955). Responses of high amplitude were recorded in the nucleus centralis lateralis of the thalamus after stimulation in the amygdala of the cat by Andy and Mukawa (1960).

The inferior thalamic peduncle or the diencephalic portion of the pedunculus thalami extracapsularis of Klingler and Gloor (1960) is a fiber system connecting the substantia innominata with the thalamus, as described early by several workers (Probst, 1898, 1901; Déjerine, 1901; Vogt, 1908; Gurdjian, 1927; Rioch, 1931). More recently, and starting from the work of Clark and

Boggon (1933), the inferior thalamic peduncle has come to be considered as a fiber system relating the ventral forebrain regions or the rostral portion of the medial forebrain bundle area with the nucleus medialis dorsalis of the thalamus (see pp. 72–74) and, consequently, it would represent the second portion of the projections from the temporal cortex and amygdala to the thalamus that have relayed in these ventral forebrain regions.

Then, the existence in the monkey of direct connections between the amygdala and the nucleus medialis dorsalis of the thalamus (Nauta, 1961) and a related pathway from the inferior temporal gyrus and the same thalamic nuclei (Whitlock and Nauta, 1956) would represent from one side the demonstration of the existence of a pathway conveying both relayed and direct fibers from the temporo-amygdaloid complex to the thalamus (the pedunculus thalami extracapsularis of Klingler and Gloor, 1960), and from the other side the nonexistence of direct amygdalo-thalamic connections in the cat would represent another additional fact supporting the idea that throughout the phylogenetic evolution the amygdaloid complex increases in importance by adding more numerous and new direct connections with other different structures.

It should be noted that I failed to demonstrate the existence in the cat (see Cat 17) of connections between the temporal convolutions and the nucleus medialis dorsalis of the thalamus.

The significance of the thalamic connections of the amygdala has been discussed under the term of an amygdalo-thalamo-orbitofrontal organization by Nauta (1962) and reference is made to this work.

Recent studies of Angeleri, Ferro-Milone, and Parigi (1964) add further evidence on the functional dependence existing between the hippocampo-amygdaloid complex and the thalamus in relation to the development of human temporal seizures.

(II) AFFERENT AMYGDALOID CONNECTIONS

In recent years increasing evidence of multiple amygdaloid afferent connections has been gained. This refers not only to the anatomic evidence of certain cortical and subcortical afferents but to the physiologic studies showing that in the amygdala evoked potentials can be elicited from all sensory modalities.

The sensory responses of the amygdala have been studied physiologically by Gérard, Marshall, and Saul (1936), Dell and Olson (1951), Dell (1952), MacLean and Delgado (1953), Machne and Segundo (1956), Dunlop (1958), and Wendt and Albe-Fessard (1962) and have been reviewed by Gloor (1960) and Gastaut and Lammers (1961). The observations of Machne and Segundo (1956) showing that various sensory impulses may converge upon a single amygdaloid cell, together with the results obtained by Gloor (1955b) on the wide modulatory activity exerted on complex autonomic, behavioral, and somatic activities of the subcortical amygdaloid projection system, allow us to assume that the amygdala may be organized either anatomically or functionally in a fashion quite similar to that of the reticular formation of the brain stem.

(1) *Olfactory connections of the amygdala*

The olfactory connections of the amygdala have been examined and discussed in conjunction with my Golgi observations (see pp. 86–94).

(2) *Temporo-amygdaloid connections*

Concerning the temporo-amygdaloid connections, reference is made to previous discussions; see p. 112, paragraph (4), and p. 113, paragraph (5). Evidence of fibers coursing rostrally through the capsula externa, entering the area amygdaloidea anterior and rostrodorsal parts of the amygdala, have been obtained from Cat 17 (see Figs. 19 and 20). In the description of this experiment I assumed that the lesion had interrupted a great volume of afferent and efferent temporal-lobe fibers, but the exact site of origin of the temporo-amygdaloid fibers cannot be determined from this material. In the monkey Whitlock and Nauta (1956) traced fibers from the inferior temporal region to the basolateral amygdaloid complex and a few to the nucleus amygdaloideus centralis. This is in disagreement with the physiologic observations of Segundo, Naquet, and Arana (1955), who reported evidence that the temporo-amygdaloid projections originate in the gyrus temporalis superior. Lammers and Lohman (1957) described the existence of fibers connecting the temporal pole with the nucleus amygdaloideus lateralis in the cat.

Taking into consideration the results obtained from Cat 17 and those discussed previously (see pp. 109–110 and p. 113, paragraph (5)), it is tentatively suggested that the amygdala plays the role of a way station providing a relay in the transamygdaloid projection pathways of the temporal lobe, that is, the ansa lenticularis or temporo-hypothalamo-mesencephalic pathway and the pedunculus thalami extracapsularis or temporo-thalamic pathway.

(3) *Thalamo-amygdaloid connections*

The anatomic existence of thalamo-amygdaloid connections was reported by Hilpert (1928), Mittelstrass (1937), and Marburg (1948). Nauta (1962) traced fiber degeneration through the pedunculus thalami extracapsularis in the monkey from the nucleus medialis dorsalis of the thalamus to rostral parts of the amygdaloid complex; the fibers follow the same course as the fibers traced by Mittelstrass (1937) through the inferior thalamic peduncle in the monkey, dog, and cat. This connection has been discussed previously (see pp. 72–74).

In the human brain Papez (1945) described a large tract of fibers entering the amygdaloid region as a continuation of the inferior thalamic peduncle via anterior perforated substance. This author did not specify the direction of conduction of these fibers but he pointed out that the fibers spread and surround the dorsal borders of the lateral, basal, and accessory basal nuclei.

Electrophysiologic evidence of direct connections between the inner part of the nucleus medialis dorsalis of the thalamus and the amygdala has been reported recently by Sager and Butkhuzi (1962).

The connections between the diffuse thalamic projection system and the amygdala are a matter of controversy. At the present time we know that very intimate functional relations do exist between the entorhino-hippocampo-amygdaloid complex and the reticular formation (Kaada,

1951; Pribram and Bagshaw, 1953; Green and Arduini, 1954; Nauta and Whitlock, 1954; Adey, Segundo, and Livingston, 1954; Arana *et al.*, 1955; Green and Adey, 1956; Adey, Sunderland, and Dunlop, 1957; Adey, Dunlop, and Sunderland, 1958; Eidelberg, White, and Brazier, 1959; Andy and Mukawa, 1960; Wendt and Albe-Fessard, 1962; and others).

Reticulo-thalamic (Nauta and Kuypers, 1958; Scheibel and Scheibel, 1958) and spino-thalamic (Mehler, Feferman, and Nauta, 1960) fibers end in and around the zone involved by the lesion in Cat 15. The fibers described in that experiment, coursing toward the amygdala via the lamina medullaris medialis–stria terminalis, would represent a connection between the intralaminar thalamic nuclei and the amygdala (Fig. 59, arrows 11, 12, and 13) and, therefore, would place the latter under the direct control of the diffuse thalamic projection system. Moreover, the degeneration traced in Cat 15 shows the existence of important connections between the intralaminar cell groups of the thalamus and the bed nucleus of the stria terminalis (Fig. 59, arrow 10) and, consequently, thence backward toward the amygdala via stria terminalis. My Golgi studies have shown the favorable position of the bed nucleus in the rat to receive abundant synaptic contacts of the thalamo-cortical fibers piercing this nucleus (see pp. 99–102).

Evoked long-latency potentials in the amygdala were obtained by Wendt and Albe-Fessard (1962) following stimulation of the nucleus centralis lateralis of the thalamus.

It is very probable that part of the fibers mediating thalamo-amygdaloid connections originate in the nucleus centrum medianum. Nauta and Whitlock (1954), in the cat, traced many degenerating fibers from a lesion in the nucleus centrum medianum to the nucleus centralis lateralis, among other thalamic nuclei, and to the head of the nucleus caudatus. McLardy (1948), Simma (1951), Powell (1952), Powell and Cowan (1955), and Johnson (1961) also reported that the nuclei centrum medianum and centralis lateralis send fibers to the nucleus caudatus. From this nucleus neurons probably project to the bed nucleus of the stria terminalis and hence toward the amygdala via the stria terminalis. The studies of Eidelberg, White, and Brazier (1959) strongly suggest that the centrum medianum is mainly involved in the nonspecific thalamic projections to the rhinencephalon.

From lesions in the habenula Mitchell (1963) has traced in the cat degenerating fibers in the stria medullaris rostrally to rostral thalamic levels, where a few degenerating fibers enter the stria terminalis and reach the amygdala. These fibers are probably similar to the fibers traced by Bürgi (1954) from the contralateral stria medullaris (see p. 108).

(4) *Hypothalamo-amygdaloid connections*

Under this heading I shall merely mention the existence of preoptico-amygdaloid fibers in the cat coursing through the substantia innominata traced by Nauta (1958), the fibers traced by Zyo, Ôki, and Ban (1963) in the rabbit from the medial forebrain bundle to the "preamygdaloid" region, and the amygdalopetal component of the longitudinal association bundle, originated in the

hypothalamic and preoptic areas, described by Powell, Cowan, and Raisman (1963) in the rat. These fibers would reciprocate the amygdalo-preoptic and amygdalo-hypothalamic fibers traced through part of the ventral amygdalofugal pathway (see pp. 73–74 and 108–110).

(5) *Orbitofrontal-amygdaloid connections.* *The problem of the sensory amygdaloid pathways*

The existence of fibers connecting the posterior orbital cortex and adjoining anterior insula and temporo-polar cortex (*région périvalléculaire* of Gastaut and Lammers 1961) with the amygdala have been recognized on the basis of neuronographic studies (Pribram, Lennox, and Dunsmore, 1950; Kaada 1951; MacLean and Pribram, 1953). Corticofugal fibers from the anterior orbital and orbito-sylvian areas have been traced by Koikegami (1963b) in the cat with the Nauta method. He found that fibers from the anterior orbital area project into the lateral division of the amygdala and those from the orbito-sylvian area project to the medial division of the amyg-

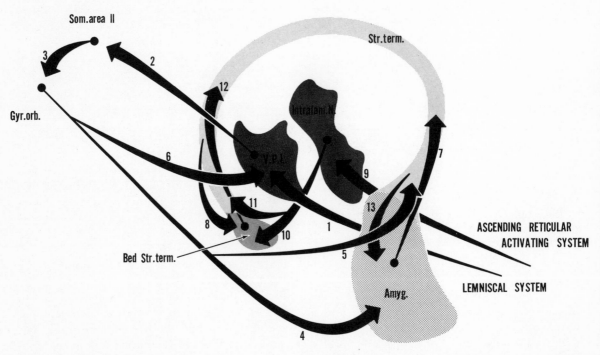

Fig. 59. Schema of the possible pathways relating the specific and nonspecific ascending brain-stem systems with the amygdaloid complex.

dala. In the monkey, Showers and Lauer (1961) have traced Marchi degeneration from the frontal and parietal opercula to the basolateral part of the homolateral amygdala, while Nauta (1962) failed to trace degeneration in this animal from the caudal orbitofrontal gyrus to the amygdala.

The observations from Cat 19 in my experimental series (confirmed in two other experiments not described, Cats 18 and 25) were discussed as representative of the connections traced from the gyrus orbitalis to the amygdaloid complex. In this experiment the lesion was situated in the lateral part of the gyrus orbitalis adjoining the gyrus ectosylvianus anterior (somatic area II). It is interesting to note that the connections traced in these experiments from the gyrus orbitalis to the amygdala furnish clear anatomic confirmation of the hypothesized pathways mediating the amygdaloid response obtained by Wendt and Albe-Fessard (1962) after stimulation in the anterior legs of the cat.

These authors have found that the pathway mediating the amygdaloid responses to contralateral anterior-limb stimulation relays in the homolateral ventralis posterolateralis thalamic nucleus (Fig. 59, arrow 1), and then in the homolateral somatic area II (Fig. 59, arrow 2), while the pathway for the response to the homolateral anterior-limb stimulation relays in the contralateral ventralis posterolateralis, then in the contralateral somatic area II, then in the homolateral somatic area II, and thence to the amygdala. They suggested such pathways on the basis of previous anatomic studies which have shown the existence of connections between the ventrobasal part of the thalamus and somatic area II (Knighton, 1950; Rose and Woolsey, 1958; Macchi, Angeleri, and Guazzi, 1959; Poggio and Mountcastle, 1960), and the interconnections between both areae somatic II (Poljak, 1927; Curtis, 1940; French, Sugar, and Chusid, 1948).

The anatomic demonstration of orbito-amygdaloid connections (Koikegami, 1963b; Valverde, 1964a) provides confirmation of the existence of the pathways described by Wendt and Albe-Fessard (1962) and agrees well with the observations reported by MacLean and Pribram (1953), which suggested that the gyrus ectosylvianus anterior

(somatic area II) links, by means of short corticocortical fibers, with the gyrus orbitalis (Fig. 59, arrow 3) and thence to the cortex praepiriformis and amygdala (Fig. 59, arrow 4). The relatively long latencies observed by Wendt and Albe-Fessard (1962) in the response of the amygdala to direct stimulation of somatic area II are in agreement with these observations.

The participation of the cortex in the mediation of somatic responses of the amygdala would represent the establishment of a new transcortical thalamo-amygdaloid circuit (Fig. 59, arrows 2, 3, and 4), whose importance in the complex behavioral pattern of the amygdaloid activity need not be emphasized. The importance of the participation of area II in the somatic responses of the amygdala is outlined by Wendt and Albe-Fessard (1962) on the resemblance of the overlapping bilateral representation of different areas of the body surface within the amygdala and on the overlapping somatotopy and bilateral responses of somatic area II. Thus it can be suggested that similar transcortical thalamo-amygdaloid pathways can be expected for the remainder of the sensory modalities through its corresponding thalamic nuclei and secondary sensory areas.

Further possibilities exist concerning the sensory pathways to the amygdala. These refer to the mediation of sensory impulses through the limbic lobe. Since the studies of Gérard, Marshall, and Saul (1936), it is known that a variety of sensory responses can be recorded in the gyrus fornicatus (Robinson and Lennox, 1951; Lennox and Madsen, 1952) and hence through the fasciculus cinguli to the amygdala (Beevor, 1891; Hilpert, 1928; Klingler, 1948), although fibers of the fasciculus cinguli entering the amygdala have been questioned by others (Gardner and Fox, 1948; Adey and Meyer, 1952; Glees et al., 1950; Showers, 1959). MacLean (1949) has emphasized the existence of connections between the limbic cortex and all the secondary sensory areas. Moreover, the participation of the limbic cortex adds still more complexity to the problem of the sensory amygdaloid pathways, suggesting that multiple and very highly complicated neuronal chains mediate the sensory influences, not only to the amygdala, but to the entire rhinencephalon.

Abbreviations Employed in Figures

A.a.	Area amygdaloidea anterior
A.b.	Nucleus amygdaloideus basalis
A.b.m.	Nucleus amygdaloideus basalis, pars magnocellularis
A.b.p.	Nucleus amygdaloideus basalis, pars parvocellularis
A.c.	Nucleus amygdaloideus centralis
Acb.	Nucleus acumbens septi
A.c.l.	Nucleus amygdaloideus centralis, pars lateralis
A.c.m.	Nucleus amygdaloideus centralis, pars medialis
A.co.	Nucleus amygdaloideus corticalis
A.D.	Nucleus anterodorsalis thalami
A.En.	Area entorhinalis
A.l.	Nucleus amygdaloideus lateralis
A.M.	Nucleus anteromedialis thalami
A.m.	Nucleus amygdaloideus medialis
Amyg.	Amygdaloid complex
An.len.	Ansa lenticularis
A.Olf.	Nucleus olfactorius anterior
A.P.M.	Area paraolfactoria medialis
A.pp.	Area praepiriformis
A.T.	Angular tract
A.V.	Nucleus anteroventralis thalami
B.C.A.	Bed nucleus of the commissura anterior
B.C.I.	Brachium conjunctivum inferior
B.Olf.	Bulbus olfactorius
B.S.t.	Bed nucleus of the stria terminalis
C.A.	Commissura anterior
C.A.p.b.	Commissura anterior, pars bulbaris
C.A.p.t.	Commissura anterior, pars temporalis
C.C.	Corpus callosum
Cd.	Nucleus caudatus
Cd.Put.	Nucleus caudatus-putamen
Cd.t.	Tail of the nucleus caudatus
C.em.	Capsula extrema
C.en.	Capsula externa
C.Fr.	Frontal cortex
Ch.	Chiasma opticum

C.I.	Capsula interna
C.im.	Capsula intermedia
C.I.p.l.	Capsula interna, posterior limb
Cl.	Claustrum
C.L.	Nucleus centralis lateralis thalami
C.M.	Nucleus centrum medianum thalami
Cng.	Cingulum
C.P.	Commissura posterior
C.Pr.	Campus prerubralis
C.S.	Colliculus superior
D.B.B.	Diagonal band of Broca
D.P.	Deep plexus of the cortex piriformis
En.	Nucleus entopeduncularis
F.D.	Fascia dentata
F.L.	Fibrillar layer of the cortex piriformis
F.l.a.	Fasciculus longitudinalis associationis
F.R.	Fissura rhinalis
Fx.	Fornix
G.C.	Gyrus cinguli
G.Cor.	Gyrus coronalis
G.E.A.	Gyrus ectosylvianus anterior
G.E.P.	Gyrus ectosylvianus posterior
gl.	Olfactory glomeruli
G.L.	Nucleus geniculatus lateralis
G.M.	Nucleus geniculatus medialis
G.Orb.	Gyrus orbitalis
G.P.	Globus pallidus
G.Pr.	Gyrus proreus
G.S.A.	Gyrus sylvianus anterior
G.Sg.	Gyrus sigmoideus
G.S.P.	Gyrus sylvianus posterior
G.Su.A.	Gyrus suprasylvianus anterior
G.Su.P.	Gyrus suprasylvianus posterior
H.A.	Hypothalamus anterior
Hb.L.	Nucleus lateralis habenulae
Hb.M.	Nucleus medialis habenulae
Hipp.	Hippocampus
H.L.	Hypothalamus lateralis
H.P.	Hypothalamus posterior

Hyp.	Hypothalamus
H1, H2	H1 and H2 fields
I.C.	Insulae of Celleja
L.D.	Nucleus lateralis dorsalis thalami
L.M.	Lemniscus medialis
L.M.M.	Lamina medullaris medialis thalami
L.M.S.	Lamina medullaris superficialis thalami
L.P.	Nucleus lateralis posterior thalami
m.c.l.	Mitral-cell layer of the olfactory bulb
M.D.	Nucleus medialis dorsalis thalami
M.F.B.	Medial forebrain bundle
Ml.	Nucleus mammillaris lateralis
Mm.	Nucleus mammillaris medialis
M.Spt.	Medial septal region
N.An.len.	Nucleus ansae lenticularis
N.C.M.	Nucleus centralis medialis thalami
N.C.P.	Nucleus commissurae posterioris
N.R.	Nucleus ruber
N.T.	Needle tract
N.T.Of.	Nucleus tractus olfactorii lateralis
P.	Nucleus posterior thalami
Ped.	Pedunculus cerebri
P.f.	Perforant fascicles of hippocampus
Pf.	Nucleus parafascicularis thalami
Pir.	Cortex piriformis
Prt.	Pretectal region
Pul.	Nucleus pulvinar thalami
Put.	Nucleus putamen
Pt.	Nucleus parataenialis thalami
P.T.D.	Pedunculus thalami dorsalis
P.T.I.	Pedunculus thalami inferior
Pv.A.	Nucleus paraventricularis anterior
Pv.H.	Nucleus paraventricularis hypothalami
R.	Nucleus reticularis thalami
Re.	Nucleus reuniens

Ret.Mes.	Reticular formation of mesencephalon
Rh.	Nucleus rhomboidalis thalami
R.L.A.	Regio limbica anterior
R.Olf.int.	Radiatio olfactoria interna
R.Per.	Regio perirhinalis
R.P.O.	Regio praeoptica
S.	Stria medullaris
Sb.	Subiculum
S.G.	Nucleus suprageniculatus
S.N.	Substantia nigra
So.	Nucleus supraopticus
Spt.	Septum
S.t.	Stria terminalis
S.t.com.	Commissural component of the stria terminalis
S.Th.	Nucleus subthalamicus
S.t.1–S.t.4	The four Johnston's components of the stria terminalis
Thal.	Thalamus
T.H.P.	Tractus habenulo-interpeduncularis
T.M.T.	Tractus mammillo-thalamicus
T.O.	Tractus opticus
T.Of.	Tractus olfactorius lateralis
Tu.Olf.	Tuberculum olfactorium
Tu.-Pir.f.	Tuberculo-piriform system of fibers
Tu.-Spt.f.	Tuberculo-septal system of fibers
V.	Ventriculus cerebri
V.A.	Nucleus ventralis anterior thalami
V.L.	Nucleus ventralis lateralis thalami
V.M.	Nucleus ventralis medialis thalami
V.P.L.	Nucleus ventralis posterolateralis thalami
V.P.M.	Nucleus ventralis posteromedialis thalami
Z.I.	Zona incerta
Z.R.	Zuckerkandl's radiation

Literature Cited

Adey, W. R. (1953), "An experimental study of the central olfactory connexions in a marsupial (*Trichosurus vulpecula*)," *Brain 76*: 311–330.

—— (1958), "Organization of the rhinencephalon," in H. Jasper *et al.*, ed., *Reticular formation of the brain* (Henry Ford Hospital Symposium; Boston: Little, Brown), pp. 621–644.

Adey, W. R., Dunlop, C. W., and Sunderland, S. (1958), "A survey of rhinencephalic interconnections with the brain stem," *J. Comp. Neurol. 110*: 173–203.

Adey, W. R., Merrillees, N. C. R., and Sunderland, S. (1956), "The entorhinal area; behavioural, evoked potential and histological studies of its interrelationships with brain-stem regions," *Brain 79*: 414–439.

Adey, W. R., and Meyer, M. (1952), "Hippocampal and hypothalamic connexions of the temporal lobe in the monkey," *Brain 75*: 358–384.

Adey, W. R., Segundo, J. P. and Livingstone, R. B. (1954), "Cortical influences on brain stem conduction," *Am. J. Physiol. 179*: 613–614.

Adey, W. R., Sunderland, S., and Dunlop, C. W. (1957), "The entorhinal area; electrophysiological studies of its interrelations with rhinencephalic structures and the brain stem," *EGG Clin. Neurophysiol. 9*: 309–324.

Ajmone-Marsan, C., and Stoll, J. (1951), "Subcortical connections of the temporal pole in relation to temporal lobe seizures," *Arch. Neurol. Psychiat. 66*: 669–686.

Albrecht, M. H., and Fernstrom, R. C. (1959), "A modified Nauta-Gygax method for human brain and spinal cord," *Stain Technol. 34*: 91–94.

Allen, W. F. (1948), "Fiber degeneration in Ammon's horn resulting from extirpations of the piriform and other cortical areas and from transection of the horn at various levels," *J. Comp. Neurol. 88*: 425–438.

Allison, A. C. (1953), "The structure of the olfactory bulb and its relationship to the olfactory pathways in the rabbit and the rat," *J. Comp. Neurol. 98*: 309–353.

Allison, A. C., and Warwick, R. T. (1949), "Quantitative observations on the olfactory system of the rabbit," *Brain 72*: 186–197.

Andy, O. J., and Mukawa, J. (1960), "Amygdaloid propagation to the brain stem (Electrophysiological study)," *EEG Clin. Neurophysiol. 12*: 333–343.

Angeleri, F., and Carreras, M. (1956), "Problemi di fisiologia dell'olfatto. I. Studio elettrofisiologico delle vie centrifughe di origine paleocorticale," *Riv. Neurobiol. 2*: 255–274.

Angeleri, F., Ferro-Milone, F., and Parigi, S. (1964), "Electrical activity and reactivity of the rhinencephalic, pararhinencephalic and thalamic structures: prolonged implantation of electrodes in man," *EEG Clin. Neurophysiol. 16*: 100–129.

Angevine, J. B., Locke, S., and Yakovlev, P. I. (1964), "Limbic nuclei of the thalamus and connections of limbic cortex. V. Thalamocortical projection of the magnocellular medial dorsal nucleus in man," *Arch. Neurol. 10*: 165–180.

Arana-Iñiguez, R., Reis, D. J., Naquet, R., and Magoun, H. W. (1955), "Propagation of amygdaloid seizures," *Acta Neurol. Lat. Amer. 1*: 109–122.

Arduini, A., and Moruzzi, G. (1953), "Sensory and thalamic synchronization in the olfactory bulb," *EEG Clin. Neurophysiol. 5*: 235–242.

Auer, J. (1953), "Some afferent connections of the hypothalamus," *Proc. Canad. Neurol. Ass. Vth Annual Meet.*, Winnipeg, June 20.

Bailey, P., Bonin, G. von, Davis, E. W., Garol, H. W., and McCulloch, W. S. (1944), "Further observations on associational pathways in the brain of *Macaca mulatta*," *J. Neuropath. Exptl. Neurol. 3*: 413–415.

Bailey, P., Bonin, G. von, Garol, H. W., and McCulloch, W. S. (1943), "Long association fibers in cerebral hemispheres of monkey and chimpanzee," *J. Neurophysiol. 6*: 129–134.

Bailey, P., Garol, H. W., and McCulloch, W. S. (1941), "Cortical origin and distribution of corpus callosum and anterior commissure in the chimpanzee (*Pan satirus*)," *J. Neurophysiol. 4*: 564–571.

Baldwin, M., Frost, L. L., and Wood, C. D. (1954), "Investigation of the primate amygdala. Movements of the face and jaws," *Neurology 4*: 586–598.

—— (1956), "Investigation of the primate amygdala. Movements of the face and jaws. 2. Effect of selective cortical ablation," *Neurology 6*: 288–293.

Ban, T., and Omukai, F. (1959), "Experimental studies on the fiber connections of the amygdaloid nuclei in the rabbit," *J. Comp. Neurol. 113*: 245–279.

Ban, T., and Zyo, K. (1962), "Experimental studies on the fibre connections of the rhinencephalon. I. Albino rat," *Med. J. Osaka Univ. 12*: 385–424.

——— (1963), "Experimental studies on the mammillary peduncle and mammillotegmental tracts in the rabbit," *Med. J. Osaka Univ. 13*: 241–270.

Bard, P. A. (1928), "A diencephalic mechanism for the expression of rage with special reference to the sympathetic nervous system," *Am. J. Physiol. 84*: 490–515.

Baumgarten, R. von, Green, J. D., and Mancia, M. (1962), "Recurrent inhibition in the olfactory bulb. II. Effects of antidromic stimulation of commissural fibers," *J. Neurophysiol. 25*: 489–500.

Beck, E., Meyer, A., and Lebeau, J. (1951), "Efferent connections of the human prefrontal region with reference to fronto-hypothalamic pathways," *J. Neurol. Neurosurg. Psychiat. 14*: 295–302.

Beevor, C. E. (1891), "On the course of the fibers of the cingulum and the posterior parts of the corpus callosum and fornix in the marmoset monkey," *Phil. Trans. Roy. Soc. (London) (B) 182*: 135–199.

Berry, C. M., Hagamen, W. D., and Hinsey, J. C. (1952), "Distribution of potentials following stimulation of olfactory bulb in cat," *J. Neurophysiol. 15*: 139–148.

Bischoff, E. (1900), "Beitrag zur Anatomie des Igelgehirnes." *Anat. Anz. 18*: 348–358.

Blackstad, T. W. (1956), "Commissural connections of the hippocampal region in the rat, with special reference to their mode of termination," *J. Comp. Neurol. 105*: 417–539.

Bonin, G. von, and Green, J. R. (1949), "Connections between orbital cortex and diencephalon in the macaque," *J. Comp. Neurol. 90*: 243–254.

Bovard, E. W., and Gloor, P. (1961), "Effect of amygdaloid lesions on plasma corticosterone response of the albino rat to emotional stress," *Experientia 17*: 521–526.

Breathnach, A. S., and Goldby, F. (1954), "The amygdaloid nuclei, hippocampus and other parts of the rhinencephalon in the porpoise (*Phocaena phocaena*)," *J. Ant. (London) 88*: 267–291.

Broca, P. (1878), "Anatomie comparée des circonvolutions cérébrales. Le grand lobe limbique et le scissure limbique dans la série des mammifères," *Rev. Anthrop. (2) 1*: 385–498.

Brockhaus, H. (1938), "Zur normalen und pathologischen Anatomie des Mandelkerngebietes," *J. Psychol. Neurol. 49*: 1–136.

Brodal, A. (1947), "The amygdaloid nucleus in the rat," *J. Comp. Neurol. 87*: 1–16.

——— (1948), "The origin of the fibers of the anterior commissure in the rat. Experimental studies," *J. Comp. Neurol. 88*: 157–205.

Brown, S., and Schäfer, E. A. (1889), "An investigation into the function of the occipital and temporal lobes of the monkey's brain," *Phil. Trans. Roy. Soc. (London) (B) 179*: 303–327.

Bucy, P. C., and Klüver, H. (1940), "Anatomic changes secondary to temporal lobectomy," *Arch. Neurol. Psychiat. 44*: 1142–1146.

Bürgi, S. (1954), "Über zwei Anteile der Stria medullaris und die Frage eines besonderen neurovegetativen Mechanismus," *Arch. Psychiat. Z. Neurol. 192*: 301–310.

Cajal, S. R. (1889), *Nuevas aplicaciones del método de Golgi* (Barcelona: J. Balmas Planas).

——— (1890), "Orígen y terminación de las fibras nerviosas olfatorias," *Gac. San. Barcelona* (October).

——— (1891), "Sobre la existencia de bifurcaciones y colaterales en los nervios sensitivos y substancia blanca del cerebro," *Gac. San. Barcelona* (April).

——— (1893), "La rétine des vertébrés," *La Cellule 9*: 119–255.

——— (1901), "La corteza olfativa del cerebro," *Trab. Lab. Invest. biol. 1*: 1–140.

——— (1911), *Histologie du système nerveux de l'homme et des vertébrés*, vol. II (Paris: Maloine); reprinted (Madrid: Instituto Cajal, 1955).

——— (1955), *Studies on the cerebral cortex (limbic structures)*, trans. by Lisbeth M. Kraft (London: Lloyd-Luke (Medical Books) Ltd.).

Cajal, S. R. and Castro, F. de (1933), *Elementos de técnica micrográfica del sistema nervioso* (Madrid: Tipografía Artística).

Calleja, C. (1893), *La región olfatoria del cerebro* (Madrid: Nicolás Moya).

Cannon, W. B. (1929), *Bodily changes in pain, hunger, fear and rage. An account of recent researches into the function of emotional excitement* (New York: Appleton).

Carreras, M., and Angeleri, F. (1956), "Problemi di fisiologia dell'olfatto. II. Una ipotesi sulla regolazione centrale dell'attivita afferente nel sistema olfattivo," *Riv. Neurobiol. 2*: 275–297.

Clark, W. E. Le Gros (1957), "Inquiries into the anatomical basis of olfactory discrimination. Ferrier Lecture," *Proc. Roy. Soc. (London) 146*: 299–319.

Clark, W. E. Le Gros, and Boggon, R. H. (1933), "On the connexions of the medial cell groups of the thalamus," *Brain 56*: 83–98.

Clark, W. E. Le Gros, and Meyer, M. (1947), "The terminal connexions of the olfactory tract in the rabbit," *Brain 70*: 304–328.

——— (1949), "Anatomical relationships between the cerebral cortex and the hypothalamus," *Brit. Med. Bull. 6*: 341–345.

Clark, W. E. Le Gros, and Warwick, R. T. (1946), "The pattern of olfactory innervation," *J. Neurol. Psychiat. 9*: 101–111.

Cowan, W. M., Guillery, R. W., and Powell, T. P. S. (1964), "The origin of the mammillary peduncle and other hypothalamic connexions from the midbrain," *J. Anat. (London) 98*: 345–363.

Cox, W. (1891), "Imprägnation des centralen Nervensystem mit Quecksilbersalzen," *Arch. Mikr. Anat.* 37: 16-21.

Cragg, B. G. (1960), "Responses of the hippocampus to stimulation of the olfactory bulb and of various afferent nerves in five mammals," *Exptl. Neurol.* 2: 547-572.

—— (1961a), "The connections of the habenula in the rabbit," *Exptl. Neurol.* 3: 388-409.

—— (1961b), "Olfactory and other afferent connections of the hippocampus in the rabbit, rat and cat," *Exptl. Neurol.* 3: 588-600.

—— (1962), "Centrifugal fibers to the retina and olfactory bulb, and composition of the supraoptic commissures in the rabbit," *Exptl. Neurol.* 5: 406-427.

Cragg, B. G., and Hamlyn, L. H. (1957), "Some commissural and septal connexions of the hippocampus in the rabbit. A combined histological and electrical study," *J. Physiol.* 135: 460-485.

Crosby, E. C. (1917), "The forebrain of *Alligator mississippiensis*," *J. Comp. Neurol.* 27: 325-402.

Crosby, E. C., and Humphrey, T. (1939), "Studies of the vertebrate telencephalon. I. The nuclear configuration of the olfactory and accessory olfactory formations and the nucleus olfactorius anterior of certain reptiles, birds and mammals," *J. Comp. Neurol.* 71: 121-213.

—— (1941), "Studies of the vertebrate telencephalon. II. The nuclear pattern of the anterior olfactory nucleus, tuberculum olfactorium and the amygdaloid complex in adult man," *J. Comp. Neurol.* 74: 309-352.

—— (1944), "Studies of the vertebrate telencephalon. III. The amygdaloid complex in the shrew (*Blarina brevicauda*)," *J. Comp. Neurol.* 81: 285-305.

Crosby, E. C., and Woodburne, R. T. (1951), "The mammalian midbrain and isthmus regions. II. The fiber connections. C. Hypothalamo-tegmental pathways," *J. Comp. Neurol.* 94: 1-32.

Curtis, H. J. (1940), "Intercortical connections of corpus callosum as indicated by evoked potentials," *J. Neurophysiol.* 3: 407-413.

Daitz, H. M., and Powell, T. P. S. (1954), "Studies of the connections of the fornix system," *J. Neurol. Neurosurg. Psychiat.* 17: 75-82.

Déjerine, J. (1901), *Anatomie des centres nerveux* (Paris: Rueff).

Dell, P. (1952), "Corrélations entre le système végétatif, et le système de la vie de relation. Mésencephale, diencéphale et cortex cérébral," *J. Physiol. (Paris)* 44: 471-557.

Dell, P., and Olson, R. (1951), "Projections 'secondaires' mésencéphaliques, diencéphaliques et amygdaliennes des afférences viscérales vagales," *C. R. Soc. Biol. (Paris)* 145: 1088-1091.

Dunlop, C. W. (1958), "Viscero-sensory and somato-sensory representation in the rhinencephalon," *EEG Clin. Neurophysiol.* 10: 297-304.

Eccles, J. C. (1961), "The mechanism of synaptic transmission," *Ergebn. Physiol.* 51: 299-430.

Eidelberg, E., White, J. C., and Brazier, M. A. B. (1959), "The hippocampal arousal pattern in rabbits," *Exptl. Neurol.* 1: 483-490.

Escolar, J. (1954), "Sobre las conexiones del complejo amigdalino," *An. Anat.* 3: 5-17.

Fernández de Molina, A., and Hunsperger, R. W. (1959), "Central representation of affective reactions in forebrain and brain stem: electrical stimulation of amygdala, stria terminalis, and adjacent structures," *J. Physiol.* 145: 251-265.

—— (1962), "Organization of the subcortical system governing defence and flight reactions in the cat," *J. Physiol.* 160: 200-213.

Fessard, A. E. (1954), "Mechanisms of nervous integration and conscious experience," in J. F. Delafresnaye, ed., *Brain mechanisms and consciousness* (Oxford: Blackwell), pp. 200-236.

Fortuyn, J. D., Hiddema, F., and Sanders-Woudstra, J. A. (1960), "A note on rhinencephalic components of the dorsal thalamus. The parataenial and dorsomedial nuclei," in A Biemond ed., *Recent neurological research* (New York: American Elsevier), pp. 46-53.

Fox, C. A. (1940), "Certain basal telencephalic centers in the cat," *J. Comp. Neurol.* 72: 1-62.

—— (1943), "The stria terminalis, longitudinal association bundle and precommissural fornix fibers in the cat," *J. Comp. Neurol.* 79: 277-295.

—— (1949), "Amygdalo-thalamic connections in *Macaca mulatta*" (abstr.), *Anat. Rec.* 103: 537-538.

Fox, C. A., Fisher, R. R., and Desalva, S. J. (1948), "The distribution of the anterior commissure in the monkey (*Macaca mulatta*)," *J. Comp. Neurol.* 89: 245-269.

Fox, C. A., McKinley, W. A., and Magoun, H. W. (1944), "An oscillographic study of olfactory system of cats," *J. Neurophysiol.* 7: 1-16.

Fox, C. A., and Schmitz, J. T. (1943), "A Marchi study of the distribution of the anterior commissure in the cat," *J. Comp. Neurol.* 79: 297-314.

French, J. D., Hernández-Peón, R., and Livingston, R. (1955), "Projections from cortex to cephalic brain stem (reticular formation) in monkey," *J. Neurophysiol.* 18: 74-95.

French, J. D., Sugar, O., and Chusid, J. G. (1948), "Corticocortical connections of the superior bank of the sylvian fissure in the monkey (*Macaca mulatta*)," *J. Neurophysiol.* 11: 185-192.

Fukuchi, S. (1952), "Comparative-anatomical studies on the amygdaloid complex in mammals, especially in Ungulata," *Folia Psychiat. Neurol. Jap.* 5: 241-262.

Ganser, S. (1882), "Vergleichend-anatomische Studien

über das Gehirn des Maulwurfs," *Morphol. Jahrb. 7*: 591–725.

Gardner, W. D., and Fox, C. A. (1948), "Degeneration of the cingulum in the monkey" (abstr.), *Anat. Rec. 100*: 663–664.

Gastaut, H., and Lammers, H. J. (1961), *Les grandes activités du rhinencéphale*, vol. I: *Anatomie du rhinencéphale* (Paris: Masson).

Gastaut, H., Naquet, R., Vigoroux, R., and Corriol, J. (1952), "Provocation de comportements émotionnels divers par stimulation rhinencéphalique chez le chat avec électrodes à demeure," *Rev. Neurol. 86*: 319–327.

Gehuchten, A. van, and Martin, L. (1891), "Le bulbe olfactif chez quelques mammifères," *La Cellule 7*: 205–237.

Gérard, R. W., Marshall, W. H., and Saul, L. J. (1936), "Electrical activity of the cat's brain," *Arch. Neurol. Psychiat. 36*: 675–735.

Glees, P., Cole, J., Whitty, C. W., and Cairns, H. (1950), "The effect of lesions in the cingular gyrus and adjacent areas in monkeys," *J. Neurol. Neurosurg. Psychiat. 13*: 178–190.

Gloor, P. (1955a), "Electrophysiological studies on the connections of the amygdaloid nucleus in the cat. Part I: The neuronal organization of the amygdaloid projection system," *EEG Clin. Neurophysiol. 7*: 223–242.

—— (1955b), "Electrophysiological studies on the connections of the amygdaloid nucleus in the cat. Part II: The electrophysiological properties of the amygdaloid projection system," *EEG Clin. Neurophysiol. 7*: 243–264.

—— (1960), "Amygdala," in J. Field *et al.*, ed., *Handbook of physiology*: sec. I, *Neurophysiology*, vol. 2 (Washington: American Physiological Society), 1395–1420

Golgi, C. (1873), "Sulla struttura della sostanza grigia del cervello," *Gazz. Med. Ital. 31*: 244–246.

—— (1875), "Sulla fina struttura dei bulbi olfattori," *Riv. sper. Freniat. 1*: 405–425.

—— (1879), "Di una nuova reazione apparentemente nera delle cellule nervose cerebrali ottenuta col bicloruro di mercurio," *Arch. Scien. Med. 3*: 1–7.

—— (1891), "Modificazione del metodo di colorazione degli elementi nervosi col bicloruro di mercurio," *Rif. Med. Napol. 7*: 193–194.

Gray, E. G., and Young, J. Z. (1964), "Electron microscopy of synaptic structure of *Octopus* brain," *J. Cell Biol. 21*: 87–103.

Gray, P. A. (1924), "The cortical lamination pattern of the opossum, *Didelphys virginiana*," *J. Comp. Neurol. 37*: 221–263.

Green, J. D., and Adey, W. R. (1956), "Electrophysiological studies of hippocampal connections and excitability," *EEG Clin. Neurophysiol. 8*: 245–262.

Green, J. D., and Arduini, A. (1954), "Hippocampal electrical activity in arousal," *J. Neurophysiol. 17*: 533–557.

Green, J. D., Mancia, M., and Baumgarten, R. von (1962), "Recurrent inhibition in the olfactory bulb. I. Effects of antidromic stimulation of the lateral olfactory tract," *J. Neurophysiol. 25*: 467–488.

Guillery, R. W. (1956), "Degeneration in the postcommissural fornix and the mammillary peduncle of the rat," *J. Anat. (London) 90*: 350–370.

—— (1957), "Degeneration in the hypothalamic connexions of the albino rat," *J. Anat. (London) 91*: 91–115.

—— (1959), "Afferent fibres to the dorso-medial thalamic nucleus in the cat," *J. Anat. (London) 93*: 403–419.

Gurdjian, E. S. (1925), "Olfactory connections in the albino rat, with special reference to the stria medullaris and the anterior commissure," *J. Comp. Neurol. 38*: 127–263.

—— (1927), "The diencephalon of the albino rat. Studies on the brain of the rat. No. 2," *J. Comp. Neurol. 43*: 1–114.

—— (1928), "The corpus striatum of the rat. Studies on the brain of the rat. No. 3," *J. Comp. Neurol. 45*: 249–281.

Hall, E. (1960), "Efferent pathways of the lateral and basal nuclei of the amygdala in the cat" (abstr.), *Anat. Rec. 136*: 205.

—— (1963), "Efferent connections of the basal and lateral nuclei of the amygdala in the cat," *Am. J. Anat. 113*: 139–151.

Heath, R. G., Monroe, R. R., and Mickle, W. A. (1955), "Stimulation of the amygdaloid nucleus in a schizophrenic patient," *Am. J. Psychiat. 111*: 862–863.

Hernández-Peón, R. (1961), "Reticular mechanisms of sensory control," in W. A. Rosenblith, ed., *Sensory communication* (Cambridge, Mass.: Technology Press), pp. 497–520.

Herrick, C. J. (1924a), "The nucleus olfactorius anterior of the opossum," *J. Comp. Neurol. 37*: 317–359.

—— (1924b), "The amphibian forebrain. II. The olfactory bulb of *Amblystoma*," *J. Comp. Neurol. 37*: 373–396.

—— (1933), "The functions of the olfactory parts of the cerebral cortex," *Proc. Nat. Acad. Scienc. 19*: 7–14.

Hess, W. R. (1949), *Das Zwischenhirn. Syndrome, Lokalisationen, Funktionen* (Basel: Benno Schwabe).

—— (1954), *Diencephalon. Autonomic and extrapyramidal functions* (New York: Grune and Stratton).

Hilpert, P. (1928), "Der Mandelkern des Menschen. I. Cytoarchitektonik und Faserverbindungen," *J. Psychol. Neurol. 36*: 44–74.

Hilton, S. M., and Zbrożyna, A. W. (1963), "Amygdaloid region for defence reactions and its efferent pathway to the brain stem," *J. Physiol. 165*: 160–173.

Hirata, Y. (1964), "Some observations on the fine structure of the synapses in the olfactory bulb of the mouse, with particular reference to the atypical synaptic configurations," *Arch. Histol. Jap. 24*: 293–302.

Honegger, J. H. (1892), "Vergleichend-anatomischen Untersuchungen über den Fornix und die zum ihm in Beziehung gebrachten Gebilde im Gehirn des Menschen und der Saugetiere," *Rec. Zool. Suisse 5*: 201–434.

Hugelin, A., Bonvallet, M., David, R., and Dell, P. (1952), "Topographie des projections centrales du système olfactif," *Rev. Neurol. 87*: 459–463.

Humphrey, T. (1936), "The telencephalon of the bat. I. The non-cortical nuclear masses and certain pertinent fiber connections," *J. Comp. Neurol. 65*: 603–711.

Jackson, J. H. (1889), "On a particular variety of epilepsy ('intellectual aura'), one case with symptoms of organic brain disease," *Brain 11*: 179–207.

Jasper, H., and Ajmone-Marsan, C. (1952), "Corticofugal projections to the brain stem," *Arch. Neurol. Psychiat. 67*: 155–171.

—— (1954), *A stereotaxic atlas of the diencephalon of the cat.* Ottawa: National Research Council of Canada.

Jeserich, M. W. (1945), "The nuclear pattern and the fiber connections of certain non-cortical areas of the telencephalon of the mink (*Mustela vison*)," *J. Comp. Neurol. 83*: 173–211.

Johnson, T. N. (1957), "The olfactory centers and connections in the cerebral hemisphere of the mole (*Scalopus aquaticus machrinus*)," *J. Comp. Neurol. 107*: 379–425.

—— (1961), "Fiber connections between the dorsal thalamus and corpus striatum in the cat," *Exptl. Neurol. 3*: 556–569.

Johnston, J. B. (1911), "The telencephalon of selachians," *J. Comp. Neurol. 21*: 1–113.

—— (1923), "Further contributions to the study of the evolution of the forebrain," *J. Comp. Neurol. 35*: 337–481.

Kaada, B. R. (1951), "Somato-motor, autonomic and electrocorticographic responses to electrical stimulation of 'rhinencephalic' and other structures in primates, cat and dogs," *Acta Physiol. Scand. 24* (Suppl. 83): 285 pp.

—— (1954), "Temporal lobe seizures," *EEG Clin. Neurophysiol.,* Suppl. IV: 235–246.

Kaada, B. R., Andersen, P., and Jansen, J. (1954), "Stimulation of the amygdaloid nuclear complex in unanesthetized cats," *Neurology 4*: 48–64.

Kappers, A. C., Huber, G. C., and Crosby, E. C. (1936), *The comparative anatomy of the nervous system of vertebrates, including man,* vol II (New York: Macmillan).

Karibe, H. (1961), "Comparative anatomical studies on the amygdaloid complex, especially in carnivora," *Acta Inst. Anat. Niigata 50*: 99–134.

Kerr, D. I. (1960), "Properties of the olfactory efferent system," *Australian. J. Exptl. Biol. Med. Sci. 38*: 29–36.

Kerr, D. I., and Hagbarth, K. E. (1955), "An investigation of olfactory centrifugal system," *J. Neurophysiol. 18*: 362–374.

Kling, A., and Hutt, P. J. (1958), "Effect of hypothalamic lesions on the amygdala syndrome in the cat," *Arch. Neurol. Psychiat. 79*: 511–517.

Klingler, J. (1940), "Makroskopische Darstellung des Faserverlaufes im Rhinencephalon," *Verh. Freie Vereinigung der Anatomen an Schweizerischen Hochschulen,* 14th Meeting (Basel).

—— (1948), "Die makroskopische Anatomie der Ammonsformation," *Denkschr. Schweiz. Naturf. Ges. 78*: 1–78.

Klingler, J., and Gloor, P. (1960), "The connections of the amygdala and of the anterior temporal cortex in the human brain," *J. Comp. Neurol. 115*: 333–369.

Klüver, H., and Bucy, P. C. (1939), "Preliminary analysis of functions of the temporal lobes in monkeys," *Arch. Neurol. Psychiat. 42*: 979–1000.

Knighton, R. S. (1950), "Thalamic relay nucleus for the second somatic sensory receiving area in the cerebral cortex of the cat," *J. Comp. Neurol. 92*: 183–191.

Koikegami, H. (1963a), "Studies on the limbic system, especially on the amygdaloid nuclear complex," *Acta Anat. Nippon. 38*: 7–8.

—— (1963b), "Amygdala and other related limbic structures; experimental studies on the anatomy and function. I. Anatomical researches with some neurophysiological observations," *Acta Med. Biol. 10*: 161–277.

Koikegami, H., and Fuse, S. (1952), "Studies on the functions and fiber connections of the amygdaloid nuclei and periamygdaloid cortex. Experiment on the respiratory movements," *Folia Psychiat. Neurol. Jap. 5*: 188–197.

Koikegami, H., Kimoto, A., and Kido, C. (1953), "Studies on the amygdaloid nuclei and periamygdaloid cortex. Experiments on the influence of their stimulation upon motility of small intestine and blood pressure," *Folia Psychiat. Neurol. Jap. 7*: 87–108.

Koikegami, H., Kushiro, H., and Kimoto, A. (1952), "Studies on the functions and fiber connections of the amygdaloid nuclei and periamygdaloid cortex. Experiments on gastro-intestinal motility and body temperature in cat," *Folia Psychiat. Neurol. Jap. 6*: 76–93.

Koikegami, H., Yamada, T., and Usui, K. (1954), "Stimulation of amygdaloid nuclei and periamygdaloid cortex with special reference to its effects on uterine movements and ovulation," *Folia Psychiat. Neurol. Jap. 8*: 7–31.

Koikegami, H., and Yoshida, V. (1953), "Pupillary dilatation induced by stimulation of amygdaloid nuclei," *Folia Psychiat. Neurol. Jap. 7*: 109–126.

Kölliker, A. (1896), *Handbuch der Gewebelehre des Menschen,* vol. II, part 1 (Leipzig: Wilhelm Engelmann).

Krieg, W. J. (1932), "The hypothalamus of the albino rat," *J. Comp. Neurol. 55*: 19–89.

Lammers, H. J., and Lohman, A. H. (1957), "Experimenteel anatomisch onderzoek naar de verbindingen van piriforme cortex en amygdalakernen bij de kat," *Ned. T. Geneesk. 101*: 1–2.

Lammers, H. J., and Magnus, O. (1955), "Étude expérimentale de la région du noyau amygdalien du chat," *Comp. Rend. Ass. Anat. XLII Réunion:* 840–844.

Lauer, E. W. (1945), "The nuclear pattern and fiber connections of certain basal telencephalic centers in the macaque," *J. Comp. Neurol. 82*: 215–254.

Lennox, M. A., and Madsen, A. (1952), "Sensory representation in olfactory brain," *Acta Physiol. Scand. 25* (Suppl. 89): 53–54.

Lewis, P. R. (1961), "The effect of varying the conditions in the Koelle technique," in *Histochemistry of cholinesterase* (Symposium Basel, 1960); *Bibl. anat. 2*: 11–20 (Basel: Karger).

Lohman, A. H. M. (1963), "The anterior olfactory lobe of the guinea pig. A descriptive and experimental anatomical study," *Acta Anat. 53* (Suppl. 49): 109 pp.

Lohman, A. H. M., and Lammers, H. J. (1963), "On the connections of the olfactory bulb and the anterior olfactory nucleus in some mammals. An experimental anatomical study," in W. Bargmann and J. P. Schadé, ed., *Progress in brain research,* vol. 3, *The rhinencephalon and related structures* (Amsterdam: Elsevier), pp. 149–162.

Loo, Y. T. (1931), "The forebrain of the opossum, *Didelphis virginiana.* II," *J. Comp. Neurol. 52*: 1–148.

Lorente de Nó, R. (1933), "Studies on the structure of the cerebral cortex. I. The area entorhinalis," *J. Psychol. Neurol. 45*: 381–438.

——— (1934), "Studies on the structure of the cerebral cortex. II. Continuation of the study of the ammonic system," *J. Psychol. Neurol. 46*: 113–177.

Löwenthal, N. (1897), "Über das Riechhirn der Säugetiere," *Festsch. z. 69 Versammlung Deutscher Naturforscher und Arzte,* Braunschweig.

Ludwig, E., and Klingler, J. (1956), *Atlas cerebri humani,* (Basel: Karger).

Macchi, G. (1951), "The ontogenetic development of the olfactory telencephalon in man," *J. Comp. Neurol. 95*: 245–305.

Macchi, G., Angeleri, F., and Guazzi, G. (1959), "Thalamo-cortical connections of the first and second somatic sensory areas in the cat, *J. Comp. Neurol. 111*: 387–405.

Machne, X., and Segundo, J. P. (1956), "Unitary responses to afferent volleys in amygdaloid complex," *J. Neurophysiol. 19*: 232–241.

McLardy, T. (1948), "Projection of the centromedian nucleus of the human thalamus," *Brain 71*: 290–303.

MacLean, P. D. (1949), "Psychosomatic disease and the 'visceral brain.' Recent developments bearing on the Papez theory of emotion," *Psychosom. Med. 11*: 338–353.

——— (1955a), "The limbic system ('visceral brain') and emotional behavior," *Arch. Neurol. Psychiat. 73*: 130–134.

——— (1955b), "The limbic system ('visceral brain') in relation to central gray and reticulum of the brain stem. Evidence of interdependence in emotional processes," *Psychosom. Med. 17*: 355–366.

——— (1958), "Contrasting functions of limbic and neocortical systems of the brain and their relevance to psychophysiological aspects of medicine," *Am. J. Med. 25*: 611–626.

——— (1960), "Psychosomatics," in J. Field *et al.,* ed., *Handbook of physiology: sec.* I, *Neurophysiology,* vol. 3. (Washington: American Physiological Society), 1723–1744.

MacLean, P. D., and Delgado, J. M. R. (1953), "Electrical and chemical stimulation of fronto-temporal portion of limbic system in the waking animal," *EEG Clin. Neurophysiol. 5*: 91–100.

MacLean, P. D., and Pribram, K. H. (1953), "Neuronographic analysis of medial and basal cerebral cortex. I. Cat," *J. Neurophysiol. 16*: 312–323.

Magnus, O., and Lammers, H. J. (1956), "The amygdaloid-nuclear complex," *Folia Psychiat. Neurol. Neurochir. Neerl. 59*: 555–582.

Mancia, M., Green, J. D., and Baumgarten, R. von (1962), "Reticular control of single neurons in the olfactory bulb," *Arch. Ital. Biol. 100*: 463–475.

Marburg, O. (1948), "The amygdaloid complex," *Confinia Neurol. 9*: 211–216.

Mehler, W. R., Feferman, M. E., and Nauta, W. J. H. (1960), "Ascending axon degeneration following anterolateral cordotomy. An experimental study in the monkey," *Brain 83*: 718–750.

Meyer, M. (1949), "A study of efferent connections of the frontal lobe in the human brain after leucotomy," *Brain 72*: 265–296.

Meyer, M., and Allison, A. C. (1949), "An experimental investigation of the connexions of the olfactory tracts in the monkey," *J. Neurol. Neurosurg. Psychiat. 12*: 274–286.

Mitchell, R. (1963), "Connections of the habenula and of the interpeduncular nucleus in the cat," *J. Comp. Neurol. 121*: 441–457.

Mittelstrass, H. (1937), "Vergleichend-anatomische Untersuchungen über den Mandelkern der Saugetiere," *Z. Anat. Entwicklg. 106*: 717–738.

Morest, D. K. (1961), "Connexions of the dorsal tegmental nucleus in rat and rabbit," *J. Anat. (London) 95*: 229–246.

Morin, F. (1950), "An experimental study of hypothalamic connections in the guinea pig," *J. Comp. Neurol. 92*: 193–214.

Münzer, E., and Wiener, H. (1902), "Das Zwischen-

und Mittelhirn des Kaninchens und die Beziehungen dieser Teile zum übrigen Zentralnervensystem, mit besonderer Berücksichtigung der Pyramidenbahn und Schleife," *Monatsschr. Psychiat. Neurol. 12*: 241–279.

Murphy, J. P., and Gellhorn, E. (1945), "Further investigations on diencephalic cortical relations and their significance for the problem of emotion," *J. Neurophysiol. 8*: 431–447.

Nauta, W. J. H. (1956), "An experimental study of the fornix system in the rat," *J. Comp. Neurol. 104*: 247–272.

—— (1958), "Hippocampal projections and related neural pathways to the midbrain in the cat," *Brain 81*: 319–340.

—— (1961), "Fibre degeneration following lesions of the amygdaloid complex in the monkey," *J. Anat. (London) 95*: 515–531.

—— (1962), "Neural associations of the amygdaloid complex in the monkey," *Brain 85*: 505–520.

Nauta, W. J. H., and Gygax, P. A. (1954), "Silver impregnation of degenerating axons in the central nervous system: a modified technique," *Stain Technol. 29*: 91–93.

Nauta, W. J. H., and Kuypers, H. G. J. M. (1958), "Some ascending pathways in the brain stem reticular formation," in H. Jasper *et al.*, ed., *Reticular formation of the brain* (Henry Ford Hospital Symposium; Boston: Little, Brown), pp. 3–30.

Nauta, W. J. H., and Valenstein, E. S. (1958), "Some projections of the amygdaloid complex in the monkey" (abstr.), *Anat. Rec. 130*: 346.

Nauta, W. J. H., and Whitlock, D. G. (1954), "An anatomical analysis of the non-specific thalamic projection system," in J. F. Delafresnaye, ed., *Brain mechanisms and consciousness*, (Oxford: Blackwell), pp. 81–116.

O'Leary, J. L. (1937), "Structure of the primary olfactory cortex of the mouse," *J. Comp. Neurol. 67*: 1-31.

Olszewski, J., and Baxter, D. (1954), *Cytoarchitecture of the human brain stem* (Philadelphia: Lippincott).

Omukai, F. (1958), "Experimental studies on fiber connection of the amygdaloid complex in the rabbit," *Acta Anat. Nippon. 33*: 499–522.

Orrego, F. (1962), "The reptilian forebrain. III. Cross connections between the olfactory bulbs and the cortical areas in the turtle," *Arch. Ital. Biol. 100*: 1–16.

Papez, J. W. (1937), "A proposed mechanism of emotion," *Arch. Neurol. Psychiat. 38*: 725–743.

—— (1945), "Fiber tracts of the amygdaloid region in the human brain, from a graphic reconstruction of fiber connections and nuclear masses" (abstr.), *Anat. Rec. 91*: 294.

Petr, R., Holden, L. B., and Jirout, J. (1949), "The efferent intercortical connections of the superficial cortex of the temporal lobe (*Macaca mulatta*)," *J. Neuropath. Exptl. Neurol. 8*: 100–103.

Phillips, C. G., Powell, T. P. S., and Shepherd, G. M. (1961), "The mitral cells of the rabbit's olfactory bulb," *J. Physiol. 156*: 26–27.

Poggio, G. F., and Mountcastle, V. B. (1960), "A study of the functional contributions of the lemniscal and spinothalamic systems to somatic sensibility. Central nervous mechanisms in pain," *Bull. Johns Hopkins Hosp. 106*: 266–316.

Poljak, S. (1927), "An experimental study of the association callosal and projection fibers of the cerebral cortex of the cat," *J. Comp. Neurol. 44*: 197–258.

Pool, J. L. (1954), "Neurophysiological symposium: The visceral brain of man," *J. Neurosurg. 11*: 45–63.

Powell, E. W. (1963), "Septal efferents revealed by axonal degeneration in the rat," *Exptl. Neurol. 8*: 406–422.

Powell, T. P. S. (1952), "Residual neurons in the human thalamus following hemidecortication," *Brain 75*: 571–584.

Powell, T. P. S., and Cowan, W. M. (1955), "An experimental study of the efferent connections of the hippocampus," *Brain 78*: 115–132.

—— (1956), "A study of thalamo-striate relations in the monkey," *Brain 79*: 364–390.

—— (1963), "Centrifugal fibres in the lateral olfactory tract," *Nature (London) 199*: 1296–1297.

Powell, T. P. S., Cowan, W. M., and Raisman, G. (1963), "Olfactory relationships of the diencephalon," *Nature (London) 199*: 710–712.

Pribram, K. H., and Bagshaw, M. (1953), "Further analysis of the temporal lobe syndrome utilizing fronto-temporal ablations," *J. Comp. Neurol. 99*: 347–377.

Pribram, K. H., Lennox, M. A., and Dunsmore, R. H. (1950), "Some connections of the orbito-fronto-temporal, limbic and hippocampal areas of *Macaca mulatta*," *J. Neurophysiol. 13*: 127–135.

Pribram, K. H., and MacLean, P. D. (1953), "Neuronographic analysis of medial and basal cerebral cortex. II. Monkey," *J. Neurophysiol. 16*: 324–340.

Probst, M. (1898), "Experimentelle Untersuchungen über das Zwischenhirn und dessen Verbindungen, besonders die songenannte Rindenschliefe," *Deutsch. Z. Nervenheil. 13*: 384–408.

—— (1901), "Zur Kenntnis des Faserverlaufes des Temporallappens, des Bulbus olfactorius, der vorderen Commissur und des Fornix nach entsprechenden Extirpations und Durchschneidungsversuchen," *Arch. Anat. Physiol., Anat. Abteil, 6*: 338–356.

Rioch, D. McK. (1931), "Studies on the diencephalon of carnivora. Part III. Certain myelinated fiber connections of the diencephalon of the dog (*Canis familiaris*), cat (*Felis domestica*) and aevisa (*Crossarchus obscurus*)," *J. Comp. Neurol. 53*: 319–388.

Robinson, F., and Lennox, M. A. (1951), "Sensory mechanisms in hippocampus, cingulate gyrus and cerebellum of cat," *Federation Proc. 10*: 110–111.

Rose, J. E., and Woolsey, C. N. (1943), "Potential changes in the olfactory brain produced by electrical stimulation of the olfactory bulb," *Federation Proc. 2*: 42.

——— (1948), "The orbitofrontal cortex and its connections with the mediodorsal nucleus in rabbit, sheep and cat," *Res. Publ. Ass. Nerv. Ment. Dis. 27*: 210–232.

——— (1958), "Cortical connections and functional organization of the thalamic auditory system of the cat," in H. F. Harlow and C. N. Woolsey, ed., *Biological and Biochemical Bases of Behavior* (Madison: University of Wisconsin Press), pp. 127–150.

Rose, M. (1929), "Cytoarchitektonischer Atlas der Grosshirnrinde der Maus," *J. Psychol. Neurol. 40*: 1–51.

Sachs, E., Brendler, S. J., and Fulton, J. F. (1949), "The orbital gyri," *Brain 72*: 227–240.

Sager, O., and Butkhuzi, S. (1962), "Electrographical study of the relationship between the dorsomedian nucleus of the thalamus and the rhinencephalon (hippocampus and amygdala)," *EEG Clin. Neurophysiol. 14*: 835–846.

Scheibel, M. E., and Scheibel, A. B. (1958), "Structural substrates for integrative patterns in the brain stem reticular core," in H. Jasper *et al.*, ed., *Reticular formation of the brain* (Henry Ford Hospital Symposium; Boston: Little, Brown), pp. 31–55.

Schreiner, L., and Kling, A. (1953), "Behavioral changes following paleocortical injury in rodents, carnivores and primates," *Federation Proc. 12*: 128.

Segundo, J. P., Naquet, R., and Arana, R. (1955), "Subcortical connections from temporal cortex of monkey," *Arch. Neurol. Psychiat. 73*: 515–524.

Shealy, C. W., and Peele, T. L. (1957), "Studies on amygdaloid nucleus of cat," *J. Neurophysiol. 20*: 125–139.

Shepherd, G. M. (1963), "Neuronal systems controlling mitral cell excitability," *J. Physiol. 168*: 101–117.

Showers, M. J. C. (1958), "Correlation of medial thalamic nuclear activity with cortical and subcortical neuronal arcs," *J. Comp. Neurol. 109*: 261–315.

——— (1959), "The cingulate gyrus: additional motor area and cortical autonomic regulator," *J. Comp. Neurol. 112*: 231–301.

Showers, M. J. C., and Lauer, E. W. (1961), "Somatovisceral motor patterns in the insula," *J. Comp. Neurol. 117*: 107–115.

Shute, C. C. D., and Lewis, P. R. (1961), "The use of cholinesterase techniques combined with operative procedures to follow nervous pathways in the brain," in *Histochemistry of cholinesterase* (Symposium Basel, 1960); *Bibl. anat. 2*: 34–49 (Basel: Karger).

Simma, K. (1951), "Zur Projektion des Centrum medianum und Nucleus parafascicularis thalami beim Menschen," *Monatsschr. Psychiat. Neurol. 122*: 32–46.

Simpson, D. A. (1952), "The efferent fibers of the hippocampus in the monkey," *J. Neurol. Neurosurg. Psychiat. 15*: 79–92.

Smith, G. E. (1910), "Some problems relating to the evolution of the brain," *Lancet 1*: 1, 147, 221.

Smith, O. C. (1930), "The corpus striatum, amygdala and stria terminalis of *Tamandua tetradactyla*," *J. Comp. Neurol. 51*: 65–127.

Snider, R. S., and Niemer, W. T. (1961), *A stereotaxic atlas of the cat brain* (Chicago: University of Chicago Press).

Sonntag, C. F., and Woollard, H. H. (1925), "A monograph of *Orycteropus afer*. II. Nervous system, sense organs and hairs," *Proc. Zool. Soc. London 2*: 1185–1235.

Sprague, J. M., and Meyer, M. (1950), "An experimental study of the fornix in the rabbit," *J. Anat. (London) 84*: 354–368.

Sprenkel, H. B. van der (1926), "Stria terminalis and amygdala in the brain of the opossum (*Didelphis virginiana*)," *J. Comp. Neurol. 42*: 211–254.

Stoll, J., Ajmone-Marsan, C., and Jasper, H. (1951), "Electrophysiological studies of subcortical connections of anterior temporal region in cat," *J. Neurophysiol. 14*: 305–316.

Tello, J. F. (1936), "Evolution, structure et connexions du corps mamillaire chez la souris blanche, avec des indications chez d'autres mammifères," *Trab. Lab. Rech. Biol. 31*: 77–142.

Thomalske, G., Klingler, J., and Woringer, E. (1957), "Über das Rhinencephalon. Physiologischer und anatomischer Überblick," *Acta Anat. 30*: 865–901.

Thomalske, G., and Woringer, E. (1957), "Die chirurgische Behandlung der herdförmigen Epilepsien unter Ausschluss der tumorös und postnatal-traumatisch bedingten," *Acta Neurochir. 5*: 223–317.

Tsai, C. (1925), "The descending tracts of the thalamus and midbrain of the opossum, *Didelphis virginiana*," *J. Comp. Neurol. 39*: 217–248.

Valenstein, E. S., and Nauta, W. J. H. (1959), "A comparison of the distribution of the fornix system in the rat, guinea pig, cat, and monkey," *J. Comp. Neurol. 113*: 337–363.

Valverde, F. (1961), "A new type of cell in the lateral reticular formation of the brain stem," *J. Comp. Neurol. 117*: 189–195.

——— (1962), "Intrinsic organization of the amygdaloid complex. A Golgi study in the mouse," *Trab. Inst. Cajal Invest. Biol. 54*: 291–314.

——— (1963a), "Studies on the forebrain of the mouse. Golgi observations," *J. Anat. (London) 97*: 157–180.

——— (1963b), "Efferent connections of the amygdala in the cat" (abstr.), *Anat. Rec. 145*: 355.

——— (1963c), "Amygdaloid projection field," in W. Bargmann and J. P. Schadé, ed., *Progress in brain research*, vol. 3, *The rhinencephalon and related structures* (Amsterdam: Elsevier), pp. 20–30.

——— (1964a), "Afferent connections of the amygdala in the cat and rat" (abstr.), *Anat. Rec. 148*: 346.

——— (1964b), "The commissura anterior, pars bulbaris" (abstr.), *Anat. Rec. 148*: 406–407.

——— (1964c), "Indirect amygdalo-hippocampal connections" (abstr.), *Anat. Rec. 148*: 407.

Valverde, F., and Sidman, R. L. (1965), "Successful Golgi impregnations in brains of mutant mice with deficient myelination" (abstr.), *Anat. Rec. 151*: 479–480.

Vogt, C. (1908), "La myéloarchitecture du thalamus du cercopithèque," *J. Psychol. Neurol. 12*: 285–324.

Vogt, O. (1898), "Sur un faisceau septo-thalamique," *C. R. Soc. Biol. 5*: 206–207.

Völsch, M. (1906), "Zur vergleichenden Anatomie des Mandelkerns und seiner Nachbargebilde. I. Teil," *Arch. Mikr. Anat. 68*: 573–683.

——— (1910), "Zur vergleichenden Anatomie des Mandelkerns und seiner Nachbargebilde. II. Teil," *Arch. Mikr. Anat. 76*: 373–523.

Votaw, C. L., and Lauer, E. W. (1963), "An afferent hippocampal fiber system in the fornix of the monkey," *J. Comp. Neurol. 121*: 195–206.

Wall, P. D., Glees, P., and Fulton, J. F. (1951), "Corticofugal connexions of posterior orbital surface in Rhesus monkey," *Brain 74*: 66–71.

Wallenberg, A. (1901), "Das basale Riechbündel des Kaninchens," *Anat. Anz. 20*: 175–187.

Ward, A. A., and McCulloch, W. S. (1947), "The projection of the frontal lobe on the hypothalamus," *J. Neurophysiol. 10*: 309–314.

Ward, J. W. (1953), "Field spread potentials of the olfactory mechanism," *Am. J. Physiol. 172*: 462–470.

Wendt, R., and Albe-Fessard, D. (1962), *Sensory responses of the amygdala, with special reference to somatic afferent pathways* (No. 107; Paris: Éditions du Centre National de la Recherche Scientifique), pp. 171–200.

Wheatley, M. D. (1944), "The hypothalamus and affective behavior in cats," *Arch. Neurol. Psychiat. 52*: 296–316.

Whitlock, D. G., and Nauta, W. J. H. (1956), "Subcortical projections from the temporal neocortex in *Macaca mulatta*," *J. Comp. Neurol. 106*: 183–212.

Yakovlev, P. I. (1959), "Patho-architectonic studies of cerebral malformations. III. Arrhinencephalie (Holotelencephalis)," *J. Neuropath. Exptl. Neurol. 18*: 22–55.

Yakovlev, P. I., and Locke, S. (1961), "Limbic nuclei of thalamus and connections of limbic cortex. III. Corticocortical connections of the anterior cingulate gyrus, the cingulum, and the subcallosal bundle in monkey," *Arch. Neurol. 5*: 364–400.

Yamamoto, C., and Iwama, K. (1961), "Arousal reaction of the olfactory bulb," *Jap. J. Physiol. 11*: 335–345.

——— (1962), "Intracellular potential recording from olfactory bulb neurons of the rabbit," *Proc. Jap. Acad. 38*: 63–67.

Yamamoto, C., Yamamoto, T., and Iwama, K. (1963), "The inhibitory systems in the olfactory bulb studied by intracellular recording," *J. Neurophysiol. 26*: 403–415.

Young, M. W. (1936), "The nuclear pattern and fiber connections of the non-cortical centers of the telencephalon of the rabbit (*Lepus cuniculus*)," *J. Comp. Neurol. 65*: 295–401.

——— (1941), "Degeneration of the fiber tracts following experimental transection of the olfactory bulb" (*abstr.*), *Anat. Rec. 79*: 65–66.

——— (1942), "Further studies on the interbulbar fibers" (abstr.), *Anat. Rec. 82*: 480.

Zuckerkandl, F. (1888), "Der Riechbündel des Ammonhorns," *Anat. Anz. 3*: 425–434.

Zyo, K., Ôki, T., and Ban, T. (1963), "Experimental studies on the medial forebrain bundle, medial longitudinal fasciculus and supraoptic decussations in the rabbit," *Med. J. Osaka Univ. 13*: 193–239.

Index

Fasciculus uncinatus, 74
Fornix, 10, 13, 22, 28, 30, 32, 43, 71, 108, 112
Fornix longus, 71

Globus pallidus, 10, 19
Golgi method, 4–9
Gyrus cinguli, 38
Gyrus coronalis, 45
Gyrus ectosylvianus anterior, 40, 45
Gyrus ectosylvianus posterior, 40
Gyrus orbitalis, 19, 38, 40, 43, 74
   see also Orbitofrontal-amygdaloid connections
Gyrus proreus, 38, 40
Gyrus rectus, 19
Gyrus sigmoideus, 38, 45
Gyrus suprasylvianus anterior, 45
Gyrus suprasylvianus posterior, 40
Gyrus sylvianus anterior, 30, 40
Gyrus sylvianus posterior, 15, 23, 25, 40, 47

Hippocampus, 19, 22, 23, 28, 32, 71
   angular tract, 23
   see also Amygdalo-hippocampal connections
Hypothalamo-amygdaloid connections, 116
Hypothalamo-thalamic connections, 72
Hypothalamus,
   importance in emotional behavior, 1
Hypothalamus anterior, 10, 22, 30
Hypothalamus lateralis, 10, 13, 22, 25, 30, 43, 73, 99, 107, 112
   see also Medial forebrain bundle
H1 and H2 fields, 43

Inferior thalamic peduncle, 19, 22, 47, 72, 113, 115
Intermediate axoned cell type, 97
Interstitial nucleus of projection pathway of temporal cortex, 72
Intralaminar nuclei of thalamus, 32, 38
   see also Amygdala, afferent connections

Lamina medullaris medialis of thalamus, 32, 35, 38
Lamina medullaris superficialis, 32, 35
Lamina zonalis, 35
Limbic midbrain region, 69
Limbic nuclei of thalamus, 35
Longitudinal association bundle. See Fasciculus longitudinalis associationis

Medial forebrain bundle, 10, 13, 22, 30, 59, 61–74, 109, 116
   ascending fibers, 69–70, 73
   bed nucleus, 65, 72
   contributions from corpus callosum, 63–65; from cortex piriformis, 63, 65, 68; from mesencephalon, 69; from septum, 65, 72; from stria terminalis, 22, 65, 72
   relations with midbrain tegmentum, 69
Mesopallium, 2
Methods
   experimental lesions, 4
   Golgi, 4–9

Methods (continued)
   Heidenhain, 4
   Nauta, 4
Mitral cells, 50–54, 56

Nuclei of amygdala. See Amygdala
Nuclei of thalamus. See Thalamus
Nucleus accumbens septi, 19, 22, 28, 30
Nucleus caudatus. See Caudate nucleus
Nucleus entopeduncularis, 22, 43
Nucleus olfactorius anterior, 54–61
Nucleus subthalamicus, 22, 43
Nucleus supraopticus, 10, 25, 28
Nucleus tractus olfactorii lateralis, 19, 94
   see also Stria terminalis

Olfactory bulb, 32, 50–54
   afferent fibers, 50, 57–59
   atypical synapses, electron microscopy, 54
   external plexiform layer, 52
   inhibition, 58–59
   interbulbar fibers, 50, 57–59
   internal granular cells, 52–54; connections, 54, 56
   internal plexiform layer, 52–54, 56
   presynaptic inhibition, 54
   relations with contralateral bulb, 56, 59
   tufted cells, 52, 56, 58
Olfactory glomeruli, 50–54
Olfactory projection area, 2–3
Orbitofrontal-amygdaloid connections, 19, 43–46, 116–117
Orbitofrontal cortex, 71

Pedunculus thalami dorsalis, 32
Periamygdaloid cortex. See Cortex piriformis
Periglomerular cells of olfactory bulb, 50
Phylogeny, 1
Piriform-amygdaloid connections. See Cortico-amygdaloid connections
Precommissural hippocampus, 63
Preoptic region, 10, 13, 15, 17, 19, 22–23, 28, 30, 43, 47, 63, 71, 73, 105, 109
   see also Fasciculus longitudinalis associationis
Pretectal region, 43
Psychomotor epilepsy, 2
Psychosomatics, 1
Putamen, 17, 19, 30, 40, 45, 47

Radiatio olfactoria interna, 13, 63
Recurrent collaterals of mitral cells, 59
Red nucleus, 43
Regio limbica anterior, 13, 15, 17, 19, 105, 108
Regio perirhinalis, 40
Regio praeoptica. See Preoptic region
Reticular formation of mesencephalon, 43
Rostral projection system of amygdala, 13, 15, 17, 19, 22, 23, 25, 30, 47, 96, 108–109, 111–112
   see also Fasciculus longitudinalis associationis

Septum, 19, 22, 28, 32, 63, 65